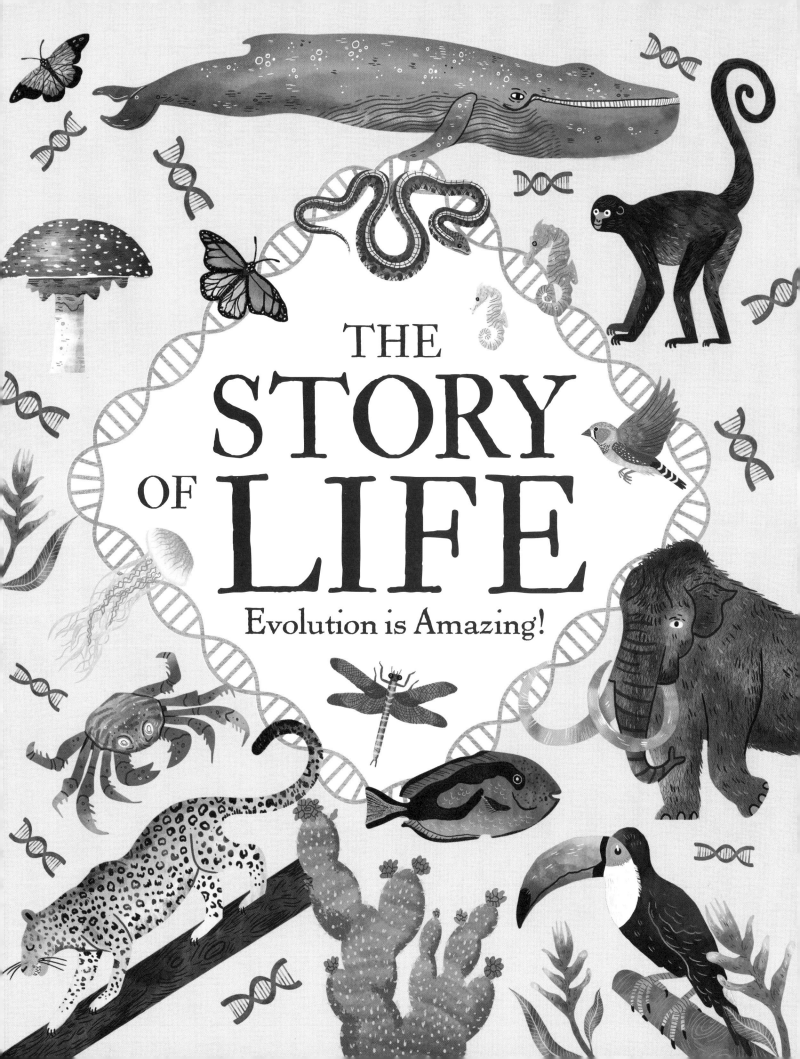

THE
STORY
OF LIFE

Evolution is Amazing!

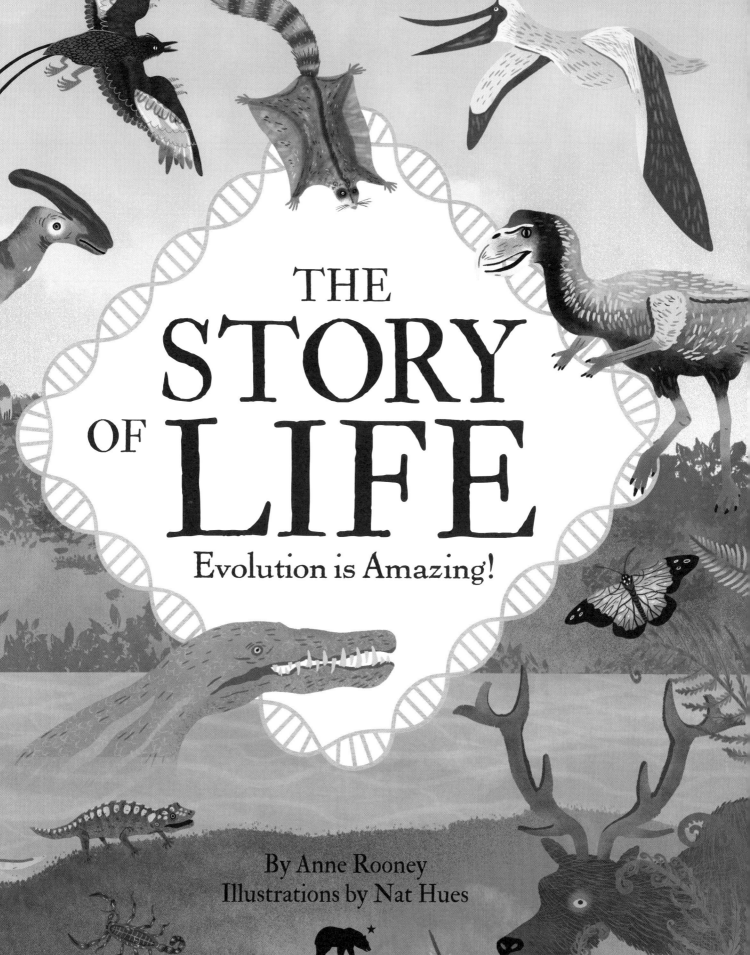

THE STORY OF LIFE

Evolution is Amazing!

By Anne Rooney

Illustrations by Nat Hues

ARCTURUS

ARCTURUS

This edition published in 2019 by Arcturus Publishing Limited
26/27 Bickels Yard, 151–153 Bermondsey Street, London SE1 3HA

ISBN: 978-1-78950-035-6
CH006564NT

Author: Dr. Anne Rooney
Consultant: Dr. Lauren Sumner-Rooney
Illustrator: Nat Hues
Art direction: Maki Ryan and Rosie Bellwood
Designer: Sally Bond
Editor: Donna Gregory
Editorial manager: Joe Harris

Supplier 29, Date 0419, Print run 7845

Printed in China

CONTENTS

Introduction

The world is teeming with life—on land, in the water, in the air, and even on the ice. There are millions of different species (types) of plants, animals, fungi, and microbes. Where have they all come from? The story of life on Earth is the story of evolution. It explains the rich variety of life as the result of living things adapting and changing to fit the available living spaces on our planet.

Asking why

The diversity of living things has fascinated people for thousands of years. Before science, we used stories and myths to explain how life came into being.

In Christian, Jewish, and Islamic tradition, God is said to have created the world and filled it with plants and animals. Other cultures have different ways of explaining the diversity of living things. The Ute people of Colorado told of a spirit called Manitou. Manitou drilled a hole in the sky, poured in rain, snow, dirt, and stones to make mountains and plains, then created trees and plants. Manitou broke up a cane and used it to make fish and animals. Manitou made birds from the leaves of the trees. When the animals began to fight, Manitou created the grizzly bear to keep order.

Manitou

In traditional stories like these, living things don't change; they are created exactly as they are now. Other stories try to explain the features of animals. The San bushmen of Namibia explain the zebra's stripes and the baboon's bare bottom with a story. The baboon would not let the zebra drink at a pool, so they fought. The zebra kicked the baboon so hard that it flew through the air and landed on its bottom on the rocks, scraping away its fur. The tired zebra staggered home, but fell into the baboon's fire and was burned in stripes across its white fur.

Stripeless zebra

Baboon

THE REAL REASON...

baboons have a bare bottom is that their bottom has a pad of tough skin with no nerves, forming a natural cushion. They can sit on rocks or stony ground and it doesn't hurt. No one is yet sure why zebras have stripes though.

From stories to science

Science takes a different approach. It looks at the world as it is and tries to explain it logically. It tests its explanations to see if they really work. One of the most important scientific explanations we have is evolution. It explains how and why living things are the way they are and accounts for life on Earth without referring to any supernatural causes.

BEING ALIVE

All life on Earth started from the simplest, tiniest organisms (living things) that appeared billions of years ago. From there, evolution has given us the dazzling variety we see around us—and an even greater variety of organisms that have been and gone in the past.

Polar bears evolved from brown bears 150,000 years ago.

Adapt or die!

Evolution is the process of adapting, so that organisms are always well-suited to conditions in the place where they live. Organisms that are not suited to their environment often die. Others will be better at finding food, a home, or a mate, and so are more likely to survive.

If a brown bear moves into a snowy landscape, it will be clearly visible against the white snow. A white bear can approach animals it wants to eat without them spotting it easily. The white bear will be more successful. If a lot of brown bears move into the snowy territory, those that are paler brown will do better than darker bears. Pale bears

will catch more food and be healthier. Being large and healthy, they will be more attractive to mates. The pale bears will breed together and have pale young. The palest young will do best. Eventually, all the successful bears will be white—white bears will have evolved from brown bears.

Changing conditions force organisms to adapt. For example, the climate might change, or a disease might emerge that kills the organism. If organisms can't adapt to meet challenges, their species might become extinct (die out).

Parent and child

All organisms inherit characteristics from their parents. You might have inherited brown eyes or curly hair. Characteristics are stored as a kind of code in your body as genes. These are tiny fragments of a chemical called DNA. They tell your body how to grow and how to act in very precise ways. Half of your genes come from your biological father and half from your biological mother, so you have a mix of characteristics from both. If one characteristic makes an organism better suited to its environment, offspring (children) with that characteristic are more likely to survive and breed, so the characteristic will become more common.

Time to change!

New characteristics turn up when mistakes happen while copying the genes from parent to child. This is called a mutation. A gene gets jumbled up, and doesn't do quite what it was supposed to do. Many mutations are bad for the organism, but others don't make much difference—you could survive just as well with violet eyes as brown eyes. Occasionally, a mutation is an improvement. It might be passed on to future generations, eventually becoming common and changing the usual form of the organism.

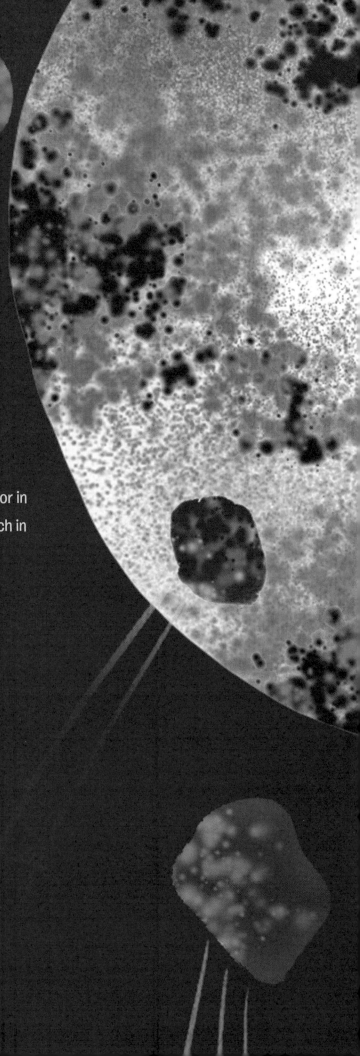

LIFE STARTS

Earth began as a lifeless lump of rock. It formed from a cloud of dust and gas whirling around the Sun 4.6 billion years ago (bya).

For 300,000 years, Earth's surface was broken apart and melted by volcanoes and by asteroids crashing into it. Finally, Earth settled into a ball of rock with oceans of liquid water. On this warmish, wet planet, life began, perhaps 4 bya.

No one knows exactly where on Earth life started, or whether it started once, a few times, or many times. It could have been in warm pools, on a hot, rocky landscape with air drenched in mist, or in the deep ocean, near vents (openings) that pour out hot water rich in minerals and gases.

A RECIPE FOR LIFE

How simple living things first formed from non-living chemicals is one of the greatest puzzles of science. Somehow, simple chemicals combined to make complex chemicals that could copy themselves. And then they needed some kind of container to separate their insides from the outside world.

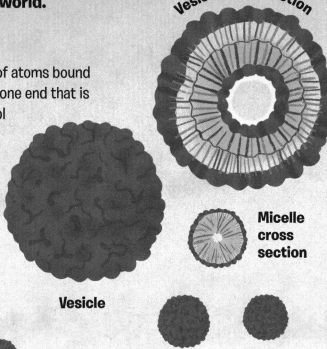

Vesicle cross section

Micelle cross section

Vesicle

Micelle

Insides and outsides

The smallest particles of matter are atoms and molecules (groups of atoms bound together). Some chemicals, called fatty acids, have molecules with one end that is attracted to water and one repelled (driven away) by water. In a pool of water, these molecules clump together with all the water-hating parts hiding in the middle and the water-loving parts on the outside. If these little balls—called micelles—collide, they can join together, making a bigger blob called a vesicle. This has two layers of molecules, and a space in the middle. The space could hold water and other chemicals and keep it apart from the rest of the environment. These vesicles became the first cells—tiny self-contained packages that are the basic building blocks of life. All living things have at least one cell—many have only one cell.

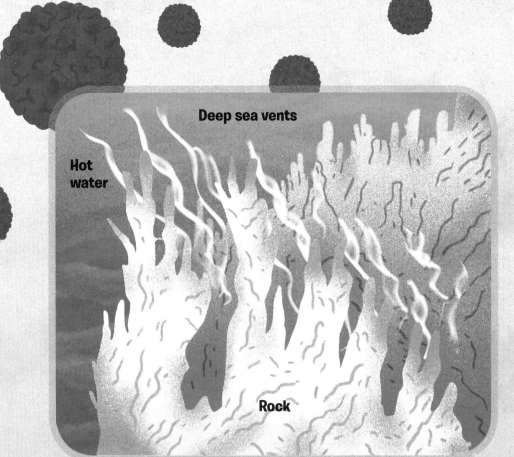

Deep sea vents

Hot water

Rock

A billion years of boring

However the first life forms began, they led dull lives. They could feed on chemicals to give themselves energy and they could reproduce, but they didn't do much else. These first organisms were called archaea. They still exist, and different kinds live everywhere, from near-boiling water in undersea vents, to deep in the mud, and in the guts of cows and termites.

Ancient organisms

Early archaea left chemical "footprints" in the oldest rocks on Earth, telling us life began at least 3.8 bya. Modern archaea give an idea of what they might have been like. They are tiny, with a length less than a tenth of the width of a human hair. Many live in scorching temperatures and "eat" chemicals to make methane (the gas we use for heating and cooking). Conditions on early Earth would have suited them just fine.

Archaea have a cell membrane—a thin barrier between chemicals inside and outside the cell. They also have a cell wall—a tougher barrier that helps them keep their shape. Inside is a liquid goop called cytoplasm, where all the chemical business of being archaea goes on.

Archaea

Cell wall

Cell membrane

Cytoplasm

ADAPTATION IN ACTION
Extremophiles

Extremophiles are organisms that can live in extremely hostile environments. Just like the first organisms on Earth, they can survive extreme heat, strong chemicals, and sometimes freezing cold or radiation. Tiny tardigrades, about 0.5 mm (0.2 in) long, have survived being taken into outer space, frozen, and dried out. The Pompeii worm lives in scalding hot-water vents in seawater full of sulfur and under great pressure.

Tardigrade

Pompeii worm

POISONED ... BY OXYGEN

Archaea were the earliest type of microbe, but others followed. The first that left fossils were bacteria. One important type were called cyanobacteria. They appeared at least 3.5 bya, and are still around today. They are simple bacteria that clump together to form a microbial mat—a carpet of living organisms. Cyanobacteria might not sound exciting but they did something that no other organism had done before—and changed the course of Earth's history in doing it.

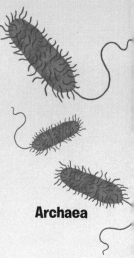

Archaea

Light lunch

Cyanobacteria use energy from sunlight to make their food from chemicals. They take in carbon dioxide and water and use them to produce a type of sugar, making oxygen as a by-product. It's a process called photosynthesis, and all green plants do it today. This way, cyanobacteria could produce 16 times as much energy as other bacteria. Their success was explosive; for everything else, it was catastrophic.

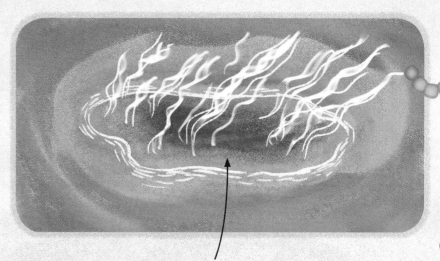

Water can be turned lots of different shades by microbes.

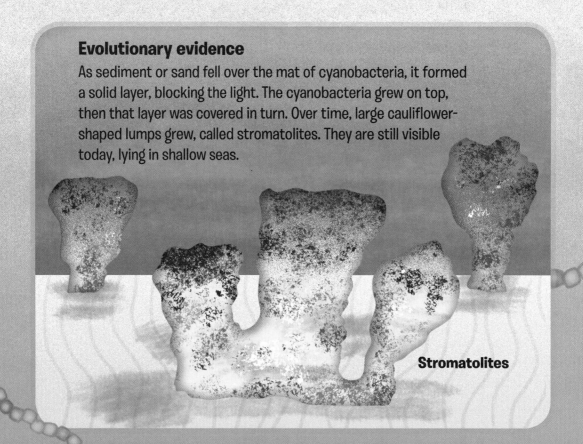

Evolutionary evidence

As sediment or sand fell over the mat of cyanobacteria, it formed a solid layer, blocking the light. The cyanobacteria grew on top, then that layer was covered in turn. Over time, large cauliflower-shaped lumps grew, called stromatolites. They are still visible today, lying in shallow seas.

Stromatolites

Too much!

The cyanobacteria produced so much oxygen that it could not dissolve in all the world's oceans. It escaped into the atmosphere, starting the oxygen-rich atmosphere we have now and that we and most other organisms depend on. The oxygen reacted with methane in the atmosphere. Methane is a powerful greenhouse gas, which traps heat near the planet. As it disappeared, heat escaped from the Earth into space and the planet became colder—much colder. Around 2.2 bya, it was winter everywhere. Earth was like a giant snowball, covered in ice from pole to pole for at least 100,000 years.

PLANTS FOREVER!

The process of photosynthesis is still used by all green plants. It now supports almost all life on Earth, as animals eat the plants, and other animals eat the plant-eaters.

Cyanobacteria

Water turned red by dying cyanobacteria.

For microbes used to a warm planet with no oxygen, the change was disastrous. The oxygen was poisonous to them, the temperature too low, and they died. This catastrophe is called the **Great Oxygenation Event**. There is evidence for it in rocks 2.3 billion years old, which show stripes of red—iron oxide—where there was so much extra oxygen that it rusted iron in the rocks.

Mass extinctions

The tiny cyanobacteria triggered the Great Oxygenation Event, causing most other organisms to die. There have been four other times when most species have died out, called mass extinction events. Some were caused by climate change, others by catastrophic events such as volcanic activity or an asteroid crashing into Earth.

TEAM PLAYERS

When the ice melted, around 2 bya, new organisms evolved to fill the gaps left by those that had died. They still had only a single cell, but it was a different kind of cell.

New cells for old

The new cells had a nucleus in which the genetic material collected. They were formed by one kind of simple cell absorbing other types. But instead of being killed, the absorbed cells carried on doing just what they did before—but from inside the other cells! They became part of the larger cells, and are called organelles (little organs). Absorbed cyanobacteria carried on making sugar using sunlight, and another type of bacteria reversed the process, releasing energy by breaking down the sugar using oxygen. Prokaryotic cells are still inside plant and animal cells today.

These new cells, called eukaryotic cells, were the ancestors of all plants, animals, and fungi. This change was one of the most important steps in the whole of evolution.

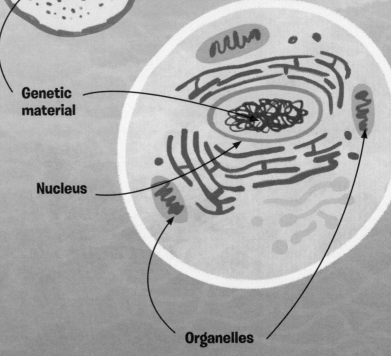

PROKARYOTIC CELL

EUKARYOTIC CELL

Genetic material

Nucleus

Organelles

ENDOSYMBIOSIS—ONE CELL ABSORBING ANOTHER

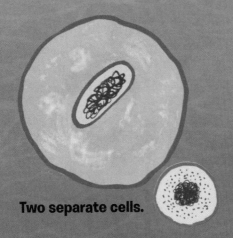

Two separate cells.

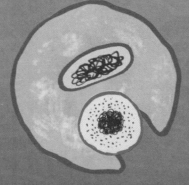

One cell engulfs another.

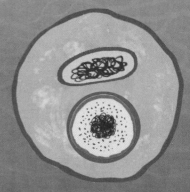

The engulfed cell becomes part of the other.

Getting together

For another billion years not much happened. Then, about 1,200 million years ago (mya), life took the next big step. *Bangiomorpha* marked a turning point in the story of life. It looks rather like modern red algae, and as far as we know, *Bangiomorpha* was the first multicelled organism. A multicelled organism can have different types of cells doing different jobs. It can have some cells that are good at holding onto a surface, and some that are good at gaining energy or making food.

Bangiomorpha
(BAN-gee-oh-morf-ah)

It takes two

Bangiomorpha did something else new—it had special cells for reproducing (making young) and became the first organism to reproduce sexually. Organisms can reproduce sexually (with two parents) or asexually (with just one parent). When an organism reproduces sexually—whether it's grass grown from seed, a penguin from an egg, or a human—it has genetic material from both of its parents. When it reproduces asexually, the parent simply splits, making an exact copy of itself (a clone). Sexual reproduction leads to variety in the offspring, and that's the best way for organisms to change quickly over generations—to evolve.

Pyrosome—
a colony of lots of tiny sea organisms that work together like a large organism.

LIVING NOW
Lichens

There are still simple organisms that live collectively and make a community without quite being a single large organism. These give us an insight into how single-celled organisms might first have come together to cooperate. Lichens are a colony of a simple organisms like cyanobacteria or algae living with filaments (fine strands) of fungus.

Lichen

INGENIOUS GENES

The genes that pass on inherited features from parent to child are sections of a long, stringy molecule of a chemical called DNA. Each string of DNA is called a chromosome. Together, all the genes completely describe an organism.

You are what you do

Your genes are a bit like a computer program for running your body. They control how your body grows and what it looks like, setting features such as the shape of your eyes and length of your fingers. They also carry instructions for the chemical processes involved in tasks like breathing and digesting food.

Humans have 23 pairs of chromosomes, each divided into hundreds of genes. Each gene tells your body how to make a particular chemical it uses, called a protein. Which proteins are produced when and where in your body controls how you grow and what your body does. There are copies of all your genes in nearly every cell of your body, kept in a special part of the cell called the nucleus.

ARE YOU A LETTUCE?

You might have seen statistics such as "you share 50 percent of your DNA with a lettuce." All organisms carry out some basic processes in the same way. Your bones grow in the same way that zebra bones grow; your cells divide in the same way lettuce cells divide. Some instructions are the same in all organisms, so some genes are the same in all organisms. The genes that you share with a lettuce control some of the processes fundamental to life on Earth.

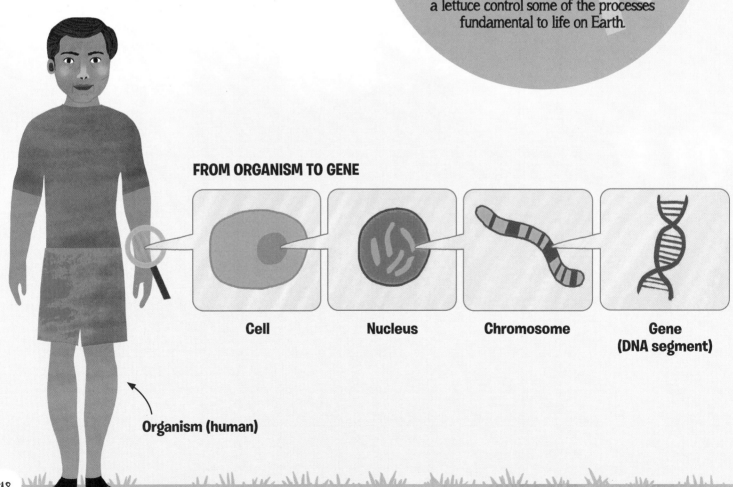

FROM ORGANISM TO GENE

Cell Nucleus Chromosome Gene (DNA segment)

Organism (human)

Making more

In all organisms, cells split to make more when the organism grows, replaces damaged or worn out cells, or reproduces asexually. First, the chromosomes copy themselves, then the two sets separate and the nucleus splits in two. The cell copies all its other contents and splits into two identical cells.

Ooops

Sometimes, something goes wrong in the copying process, making a mutation. If the organism survives, it can pass on the mutation to future generations. For organisms that reproduce asexually, mutation is the only way species can change.

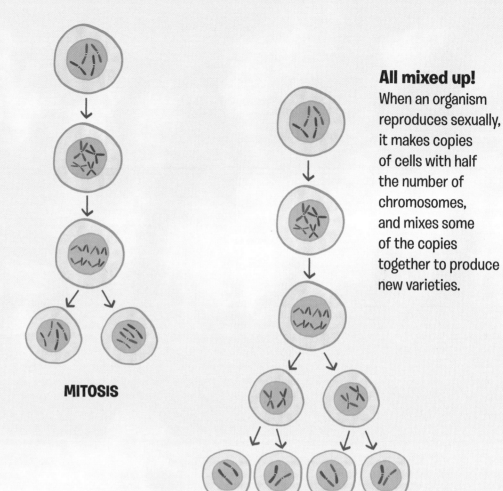

MITOSIS

MEIOSIS

All mixed up!

When an organism reproduces sexually, it makes copies of cells with half the number of chromosomes, and mixes some of the copies together to produce new varieties.

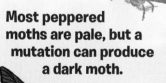

Most peppered moths are pale, but a mutation can produce a dark moth.

Which gene wins?

Many genes have variants called alleles. Maybe an allele can produce yellow petals or red petals, for instance. What happens if an organism inherits one allele for a red and one for yellow? Sometimes the result will be a blend of the two—striped petals, perhaps. Often, one gene wins. This is called the dominant allele. If yellow is dominant, the plant will always have yellow petals unless it has two alleles for red petals.

Parent

Offspring

Parent

SNOWBALLS AND SPONGES

Even when multicelled organisms got started, progress was slow. It took more than half a billion years, from 1,200 million to 600 mya, to produce something as complex as a sponge!

Simple beginnings

Sponges are very simple organisms that still live in seas all around the world. They have four different types of cells that between them help the sponge attach itself to a rocky base; hold the body in shape; waft water through holes in its body; filter out and digest tiny particles of food; and reproduce. The oldest known sponges lived 500–600 mya.

The oldest fossil sponge is less than 1 mm (1/16 in) across.

Snowball Earth

Just as the Earth turned into a giant snowball after the Great Oxygenation Event, so it did again around some time between 715 and 635 mya. Living things could only survive beneath the ice sheet and many types would have died out.

But when the world warmed again, the surviving organisms spread out and diversified, evolving into different forms. Larger and more complex organisms appeared. Everything still lived in the sea, but there were important new departures. Plants and animals became more distinct, and animals split into two main groups, which would eventually become arthropods (including insects, spiders, crabs, and scorpions) and vertebrates (animals with a backbone).

Plant or animal?

It's hard to tell whether some of the organisms that evolved were plants or animals. *Charnia* had a circle at the bottom anchoring it to the seabed, and the ridged top part moved with the water currents. It might have been a colony of tiny organisms living together. Modern sea pens have a stem and anchor part made of one organism and lots of tiny organisms making up the frond. Individuals do different tasks, such as feeding, or reproducing. Perhaps *Charnia* was similar.

Dickinsonia (565–540 mya) was the first animal. Up to 1 m (just over 3 ft) long, and had a ribbed surface. No one knows which end was the head, if it even had a head.

Charnia (char-NEE-ah)

Fold-in-half-able organisms

The organisms that appeared after Snowball Earth were the first to be symmetrical—to have two halves that could exactly mirror each other. You, and most other types of animal, have this sort of symmetry, called bilateral symmetry. But some have radial symmetry—they look as if one portion has been repeated, rotating around a central point, usually five or more times. Starfish and sea urchins have radial symmetry.

Dickinsonia
(DIH-kin-SOH-nee-ah)

Tribrachidium
(TRY-brah-KIH-dee-um)

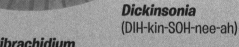

EVOLUTIONARY PUZZLES
Tribrachidium

Tribrachidium was really unusual. It had radial symmetry, but made by repeating one part three times. It's one of the only known organisms with three-fold symmetry. No one knows what type of organism it was—or even if its fossils are just part of a larger organism.

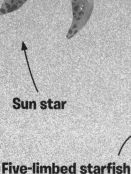

Sun star

Five-limbed starfish

21

THE TREE OF LIFE

With millions of different types of organisms both alive and extinct, scientists need a way to group them together so that it is easier to talk about them, study them and compare them. But classification doesn't just usefully divide organisms into groups. It also tries to reflect the path of their evolution, showing how they are related to each other in a gigantic family tree.

Top down

Archaea and bacteria were the first living things to evolve. Eukaryotes followed when some of the others combined, giving three groups of microorganisms. This is the top level of grouping, called domains. Everything else has evolved from these three domains.

Kingdoms of beasts and plants

The most complex organisms—all plants, animals, and fungi—are in the domain eukaryotes. "Animal" and "plant" are examples of kingdoms within the domain eukaryote. Kingdoms are the next level of classification. Animals are then divided again and again into smaller and smaller categories.

LIFE

DOMAIN Bacteria

DOMAIN Eukarya

DOMAIN Archaea

KINGDOM Animalia

KINGDOM Plantae

KINGDOM Fungi

KINGDOM Protista

PHYLUM Chordata

Only the final level defines a single type of animal. So "Animals" includes insects and worms as well as animals with backbones. "Animals with backbones" includes fish, crocodiles, and birds. "Mammals" includes such varied animals as giraffes, mice, and bears. "Cetaceans" also includes whales and porpoises.

SUBPHYLUM Vertebrata

CLASS Mammalia

ORDER Artiodactyla

ORDER Cetacea

FAMILY Delphinidae

GENUS Tursiops

SPECIES truncatus

Woolly mammoth from mainland

MULTIPLE MAMMOTHS

The fact that one organism has evolved from another doesn't mean that there are no more of the original organism. About 6,000 years ago, Wrangel Island became cut off from Siberia by the sea. A group of 500–1,000 woolly mammoths lived on the island. They evolved separately from the mammoths on the mainland, and eventually grew to have smoother, pale fur. The mammoths on the mainland still existed to start with, though they became extinct long before the island mammoths died out.

Woolly mammoth from Wrangel Island

Shared ancestors

The huge variety of organisms alive today all evolved from a small number of ancestors. Scientists try to work backward through evolution to find common ancestors—organisms from which several other types have evolved. They draw tree diagrams to show how different types of organism are related. The point where two lines separate (where the tree branches), is where two organisms that share a common ancestor have evolved in different ways. Dogs and timber wolves had a common ancestor from which they both evolved. That common ancestor also had a common ancestor with Himalayan wolves, and that one had a common ancestor with coyotes.

Dog

Timber wolf

Himalayan wolf

Coyote

23

LIFE EXPLODES

Once life had got going with larger organisms, evolution exploded dramatically. Around 540 mya, lots of different body types emerged very quickly. This is called the Cambrian Explosion. It's the point at which complex plants and animals came about, and became clearly different as plants or animals.

Burgess Shale

On a mountainside in Canada, Charles and Helena Walcott discovered in 1909 what turned out to be a huge collection of fossils that provide a window into the distant past. The Burgess Shale contains thousands of fossilized creatures in a layer of flaky rock. It's unusual in that the fossils are of soft-bodied organisms; usually, fossils preserve just the hard parts like bones and teeth.

Trilobites (TRY-loh-bytes) first appeared around 540 mya. Their bodies were divided into segments, and they had jointed legs.

AGE OLD

People who study the Earth in the distant past divide time into named geological ages. The time up to the Cambrian Explosion is called the Proterozoic, ending at 541 mya. Everything since is the Phanerozoic.

Anomalocaris (ah-NOM-ah-lo-KAR-iss) was a top predator—nothing hunted it. An early arthropod that grew up to 1 m (3 ft) long, it would have rippled its segments through the sea as it hunted with its spiky mouthparts for the smaller, softer creatures around it.

Possibly related to later mollusks like limpets, **Wiwaxia** (wee-WAX-ee-ah) lived on the seabed, probably scraping algae from the rocks with rows of rear-facing, conical teeth underneath it.

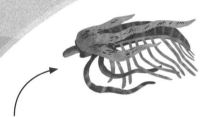

Marrella (mah-RELL-ah) looked a bit like a lacy crab, and might have been an early type of arthropod. It had two pairs of long spines stretching backward from the head, and 20 body segments, each with a pair of legs.

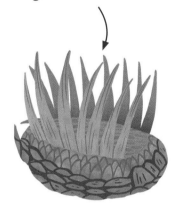

The weird-looking early arthropod **Opabinia** (OH-pah-BIN-ee-ah) had five eyes and a long, grasping organ with spines that looked rather like teeth. It used this to grab food. and pass it up to its mouth.

Another animal that could be an early arthropod is **Aysheaia** (eye-SHAY-ah). With its many short legs underneath, it looked a bit like a caterpillar, but with a savage-looking mouth surrounded by spikes.

Hallucigenia (hal-LUCE-ee-GEN-ee-ah) walked over the seabed on 7 or 8 pairs of fleshy limbs, each with a pair of claws. It had a rounded head at one end, and two rows of spikes along its back.

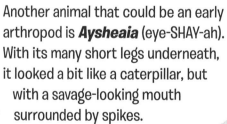

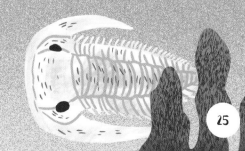

25

EYES WIDE OPEN

For the first 3.5 billion years of life on Earth, nothing could see clearly. Then, with the evolution of eyes and proper vision about 540 mya, animals could see where they were going and see each other. They could see predators chasing them and see prey they wanted to eat. Eyes are useful to animals that can move, and they have evolved independently lots of times.

Light and dark

Being able to tell light from dark is easily achieved using chemicals sensitive to light. Even plants grow toward the light. Sunlight acts on a chemical in the plant and directs its growth. But we don't think of plants as being able to see. Many bacteria can move toward the light because they have a simple "eyespot," which contains a chemical sensitive to light.

Some animals have very simple types of "eyes" that are groups of light-sensitive cells. It's enough for the animal to be able move to a shady area where it might be hidden from predators that want to eat it, or protect it from drying out in hot sunlight—or to move to a light area where it will find food. Modern animals that can do this include starfish and simple worms.

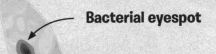
Bacterial eyespot

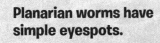
Planarian worms have simple eyespots.

USE IT OR LOSE IT

Many of the animals of the Burgess Shale had eyes, but the plants didn't. There would be no point in plants evolving eyes, as they can't use the information they would gain from sight—they can't move away from a predator. Animals that live in the dark gain nothing by having eyes and often lose their eyes and invest more in other senses like taste or touch instead. Eyes take effort to grow and can be easily hurt. Any such animal born with smaller eyes or no eyes has an advantage—it has one less complex, vulnerable part to go wrong. Organisms lose other body parts or functions that are no longer useful when their environment changes, too.

Starfish have eyespots at the end of each arm.

EVOLUTION OF EYES

Pigment spot

Optic nerve

Pigment cup

Pinhole eye

Primitive enclosed eye

Primitive lensed eye

Complex eye

Dragonflies have compound eyes.

A dragonfly's compound eye.

Seeing the light

More advanced eyes developed in the Cambrian Explosion. *Anomalocaris* had eyes with lots of tiny lenses—up to 16,000 in each eye. This type of eye is called a compound eye and insects still have them. Trilobites had crescent-shaped compound eyes. Dragonflies have compound eyes with 28,000 lenses each.

Trilobites had compound eyes.

Crescent-shaped compound eye.

All mammals, including this llama, have complex eyes.

What big eyes you have!

Our eyes, and those of lots of other animals, have a lens. This changes shape to focus light at the back of the eye where a nerve carries information to the brain. The brain builds a complex picture from the light coming into the eye—it constructs what we see, giving us vision.

Vision might have driven evolution, with animals evolving better, faster ways of moving, ways of hiding from predators, and ways of attracting mates that depend on being able to see.

27

ONTO LAND

For around 3.5 billion years, everything alive lived in the sea. Then, around 450 mya, the first living things began to explore the bare rock of the coastlines.

The first explorers were arthropods—hard-bodied, jointed creatures such as scorpions and centipedes. They were followed by plants. At first, only algae, lichens, and mats of microbes coated the rocks near the water. When these dried out and died, microbes broke up the dead material, along with waste from arthropods, making the first pockets of soil where plants took root. Early mosses grew in the damp and shady areas, then liverworts spread alongside them.

As plants grew and spread, so did animals. The arthropods were joined by strange, wet, slippery creatures as fish struggled up the mud-banks and beaches. Starting from the sea-bordered edges, evolution filled the land with animals and plants. Within 200 million years, forests of giant trees soared upward, reptiles the size of a hippopotamus crashed their way through them, and insects as large as birds hovered between the trees.

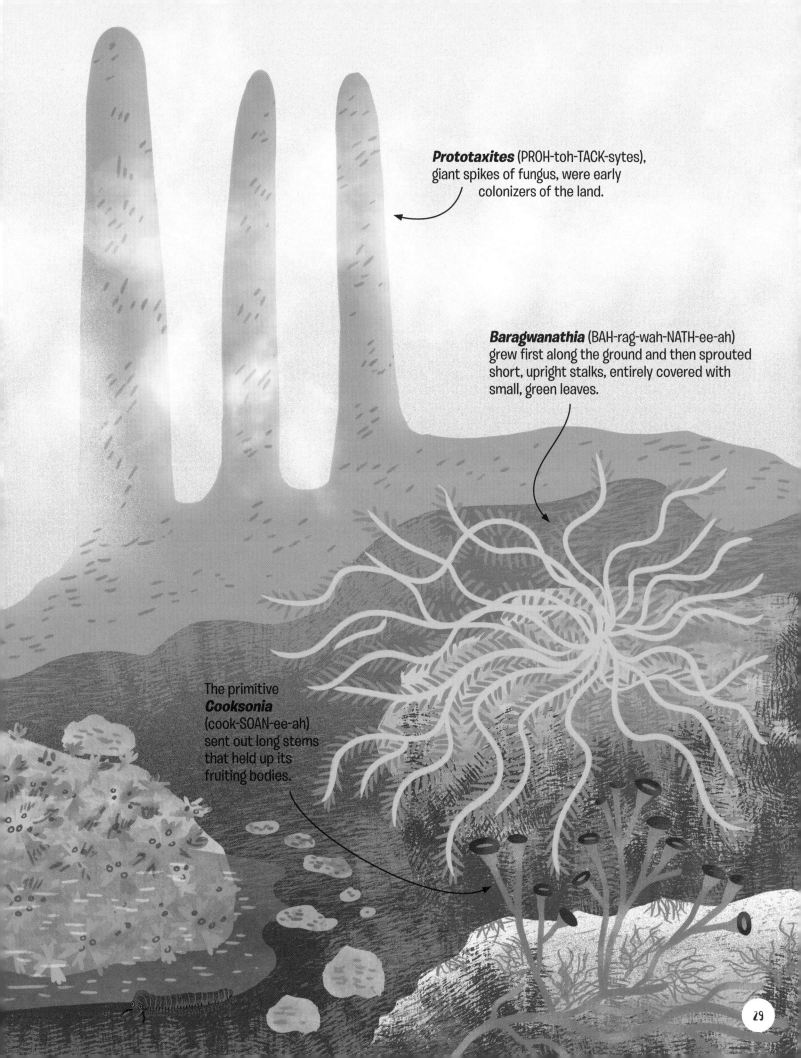

Prototaxites (PROH-toh-TACK-sytes), giant spikes of fungus, were early colonizers of the land.

Baragwanathia (BAH-rag-wah-NATH-ee-ah) grew first along the ground and then sprouted short, upright stalks, entirely covered with small, green leaves.

The primitive **Cooksonia** (cook-SOAN-ee-ah) sent out long stems that held up its fruiting bodies.

THE MOTHER OF ALL INSECTS

Today, we often overlook arthropods, but 500–400 mya, they ruled the world. Arthropods include all the jointed creatures with a hard outer skeleton, called an exoskeleton. Today's arthropods are spiders, scorpions, crabs, shrimp, millipedes, and all insects. They make up more than 75 percent of animal species.

Worms and legs

The first arthropods evolved 530 mya from worms with a body divided into segments, called annelid worms. It's quite a short step from the segments of a worm-body to a more complicated arrangement of segments.

A soft and squashy worm is vulnerable, so there was an evolutionary advantage in gaining a hard outside. But as its outside became hard, the animal needed flexible joints, like hinges, so that it could bend and move. In the early arthropods—and in millipedes and centipedes today—one body segment is repeated lots of times between the head and tail ends. More complicated arthropods have developed different, specialized body segments.

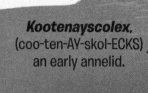

Kootenayscolex, (coo-ten-AY-skol-ECKS) an early annelid.

Earthworm, a modern annelid.

Trilobites rule the seas

Of all the early arthropods, the trilobites were most successful. They swam and crawled through all the world's seas in at least 4,000 species, ranging in size from just 6 mm to 60 cm (¼ in to 24 in). They had a head area, a body with legs or paddles, and a tail end. They were symmetrical left to right, had a top and bottom, and had eyes—they were fully equipped with all the latest and best advantages an animal could have 500 mya. Despite their success, all trilobites were wiped out 252 mya.

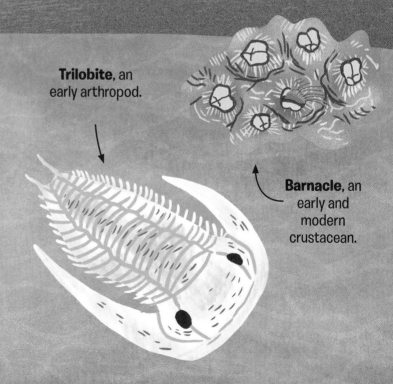

Trilobite, an early arthropod.

Barnacle, an early and modern crustacean.

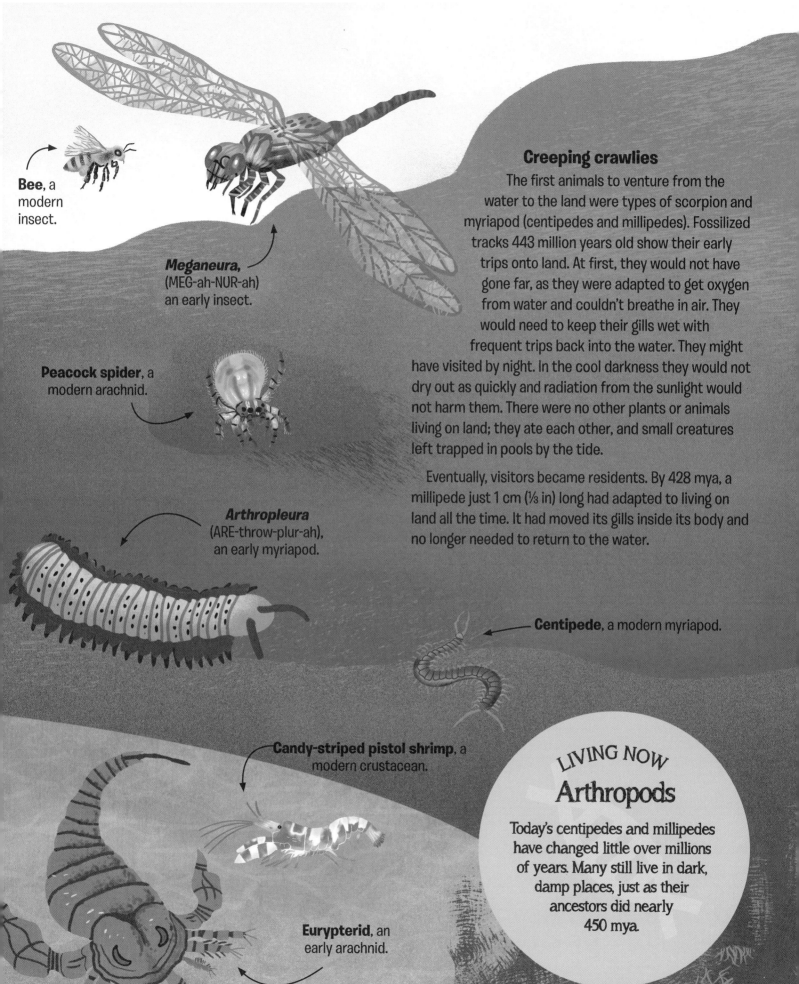

Bee, a modern insect.

Meganeura, (MEG-ah-NUR-ah) an early insect.

Peacock spider, a modern arachnid.

Arthropleura (ARE-throw-plur-ah), an early myriapod.

Centipede, a modern myriapod.

Candy-striped pistol shrimp, a modern crustacean.

Eurypterid, an early arachnid.

Creeping crawlies

The first animals to venture from the water to the land were types of scorpion and myriapod (centipedes and millipedes). Fossilized tracks 443 million years old show their early trips onto land. At first, they would not have gone far, as they were adapted to get oxygen from water and couldn't breathe in air. They would need to keep their gills wet with frequent trips back into the water. They might have visited by night. In the cool darkness they would not dry out as quickly and radiation from the sunlight would not harm them. There were no other plants or animals living on land; they ate each other, and small creatures left trapped in pools by the tide.

Eventually, visitors became residents. By 428 mya, a millipede just 1 cm (⅓ in) long had adapted to living on land all the time. It had moved its gills inside its body and no longer needed to return to the water.

LIVING NOW
Arthropods

Today's centipedes and millipedes have changed little over millions of years. Many still live in dark, damp places, just as their ancestors did nearly 450 mya.

31

LIFE UNDER THE OCEAN WAVE

While a few arthropods explored land, most life flourished in the sea. Among the sponges, trilobites, and horseshoe crabs swam jellyfish, and early squid—including ammonites in their shells. Starfish and sea urchins crawled over the seabed, and the first, strange, fish appeared.

Dunkleosteus, 382–358 mya

The fearsome giant *Dunkleosteus* (DUN-kel-os-tee-uss) grew up to 6 m (20 ft) long.

Something fishy

The earliest fish evolved from creatures like *Haikouichthys* (HY-koo-ICK-thiss), common 525 mya. Just 2.5 cm (1 in) long, *Haikouichthys* had a backbone, and moved through the water with a wrap-around fin that went from the top of its body, around the tail, and finished underneath.

The first true fish were eel-shaped, with a ring of teeth at the end. They had no jaws, which limited their feeding opportunities. They had tough cartilage in place of bones.

Ammonite

Drepanaspis, 400 mya

Drepanaspis (DREH-pan-asp-iss) lay at the bottom of the sea, feeding off the seabed. Its hard, bony plates protected it from anything trying to attack it from above.

Curled and uncurled

Among the fish swam ammonites, a kind of squid in a shell. Some were tightly curled spirals, but others were partly curled or even nearly straight. They evolved from straight-shelled relatives around 408 mya and survived until 66 mya. Ammonites evolved into many different species, each of which only lasted a few million years. Their fossils help scientists to date rock layers and so other fossils found in the same place.

Ammonite

EVOLUTIONARY PUZZLES
Stethacanthus

An early shark, *Stethacanthus*, had a very strange feature that no modern shark shares—a bony blob on its head covered with tiny mini-teeth, and a patch of more mini-teeth on its nose. No one knows what these were for.

Stethacanthus,
(STETH-ah-CAN-thuss)
323 mya

The jaws of victory

When the ancestors of modern fish evolved around 430 mya from simple jawless fish, they rapidly took over. They developed hinged jaws, which meant they could bite, and they grew paired fins, which gave them greater control over their movements.

The Devonian period, from 417–354 mya, has been called the age of fish. It saw the early ancestors of lobe-finned fish, sharks, rays, and the strange, placoderms covered with tough, bony plates.

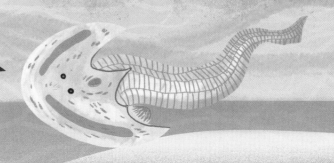

Zenaspis, 400 mya
Another bottom-feeder, *Zenaspis* (ZEN-asp-iss) had a strange, arrow-shaped head.

FEET FOR FISH

Our own ancestors were late arrivals on land. We—and all other four-legged animals—evolved from fish that struggled onto land and adapted to breathing air, around 375 mya.

Fish that can drown

Fish usually live in the water and breathe through their gills. As water flows through the mouth and out over the gills, oxygen passes into the fish's blood. But around 400 mya, lungfish started taking gulps of air at the water's surface.

Lungfish alive today can survive rivers drying out, as they can bury themselves in mud and breathe air when other fish would die. They even drown if kept only in water, where they can't breathe air.

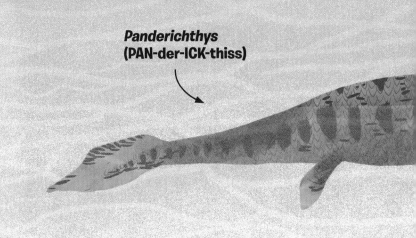

Panderichthys
(PAN-der-ICK-thiss)

Acanthostega
(ah-CAN-thoh-STAY-gah)

Crawling from the swamp

Lungfish next evolved strong fins that could prop up their body when on the ground. Fish like this hauled themselves from the rivers and up the muddy banks. Amphibians evolved from these "fishapods"— fish with feet.

The most famous fishapod is *Tiktaalik*. Growing up to 2.5 m (8 ft) long, *Tiktaalik* had a head like a crocodile's, a neck it could move (fish don't have a neck) and a strong pelvis (hip area). Fish typically have a narrow, weak pelvis, but land-going animals need strong back legs. *Tiktaalik* shows that the hips changed before fish-like animals left the water for good.

Tiktaalik (tick-TAH-lick)

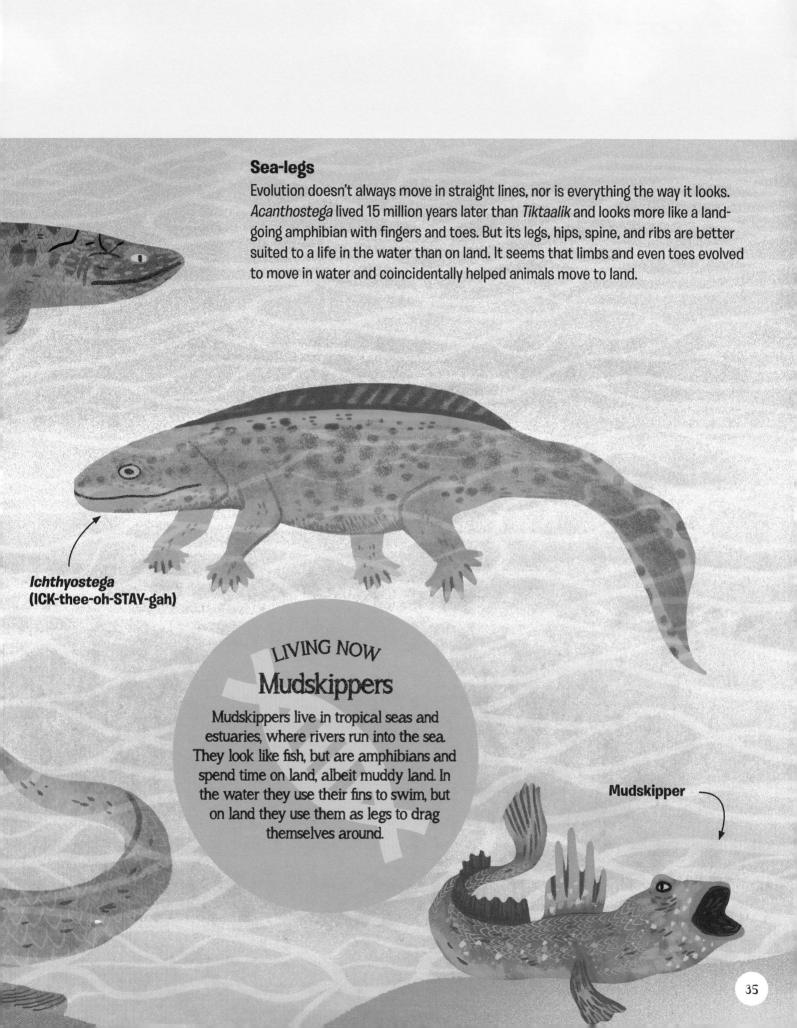

Sea-legs

Evolution doesn't always move in straight lines, nor is everything the way it looks. *Acanthostega* lived 15 million years later than *Tiktaalik* and looks more like a land-going amphibian with fingers and toes. But its legs, hips, spine, and ribs are better suited to a life in the water than on land. It seems that limbs and even toes evolved to move in water and coincidentally helped animals move to land.

***Ichthyostega*
(ICK-thee-oh-STAY-gah)**

LIVING NOW
Mudskippers

Mudskippers live in tropical seas and estuaries, where rivers run into the sea. They look like fish, but are amphibians and spend time on land, albeit muddy land. In the water they use their fins to swim, but on land they use them as legs to drag themselves around.

Mudskipper

ADAPTING TO AIR

The fish that crawled from the water adapted to living more of their lives on land and breathing air all the time, at least as adults. They evolved into the first amphibians. Even today, amphibians lay their eggs in water, and their young start with a larval stage that lives in the water and has gills.

In and out of the water

Pederpes adapted to living on land around 348 mya. About 1 m (3 ft) long, it had a mix of fishy and amphibian characteristics. Its forward-facing feet with toes were a change from the fishapods' small limbs jutting out sideways. Unlike the flat-headed fishapods, *Pederpes* had a tall, narrow head. Its ears were better adapted to hearing underwater than in air, so it possibly still hunted for food in the water.

Pederpes
(PED-er-peez)

Most amphibians are carnivores—they eat other animals—at least as adults. Inland, the early amphibians ate arthropods and some ate smaller amphibians. They didn't wander far inland, as they still had to keep their skin moist and lay their eggs in water.

Amphibamus (am-PHI-bah-muss), just 12 cm (5 in) long, was similar to modern amphibians such as frogs and salamanders. It had tiny, spiked, hinged teeth, and possibly had a larval stage a bit like a tadpole, with a long, finned tail.

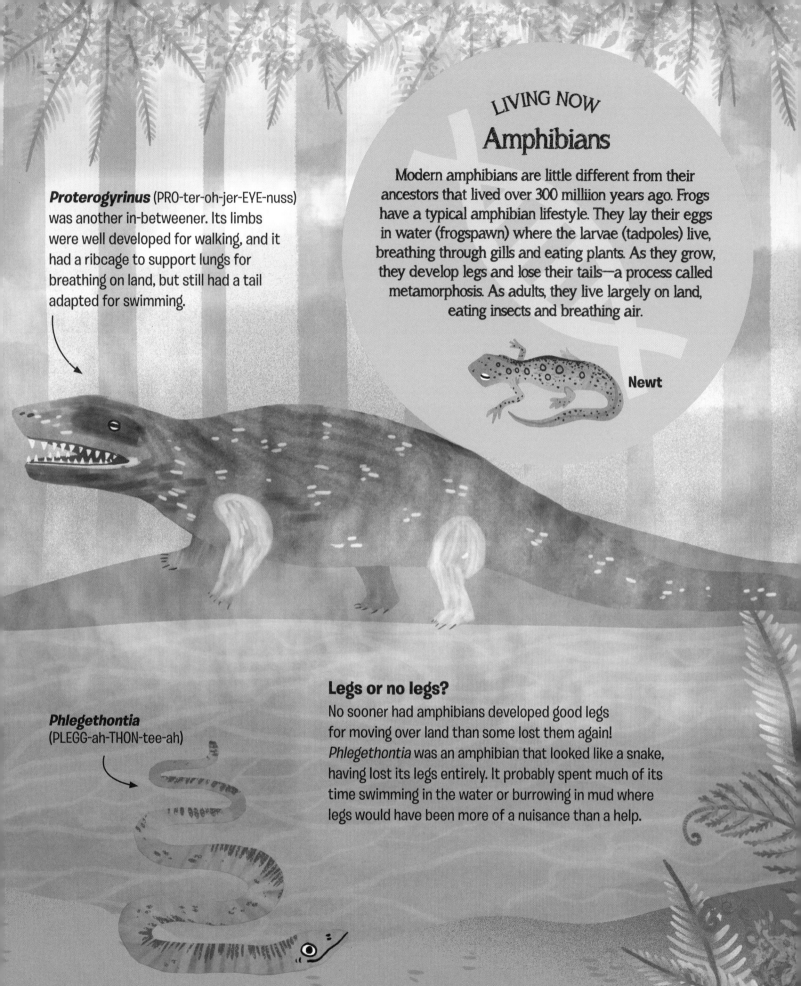

Proterogyrinus (PRO-ter-oh-jer-EYE-nuss) was another in-betweener. Its limbs were well developed for walking, and it had a ribcage to support lungs for breathing on land, but still had a tail adapted for swimming.

Amphibians

Modern amphibians are little different from their ancestors that lived over 300 milliion years ago. Frogs have a typical amphibian lifestyle. They lay their eggs in water (frogspawn) where the larvae (tadpoles) live, breathing through gills and eating plants. As they grow, they develop legs and lose their tails—a process called metamorphosis. As adults, they live largely on land, eating insects and breathing air.

Newt

Phlegethontia
(PLEGG-ah-THON-tee-ah)

Legs or no legs?

No sooner had amphibians developed good legs for moving over land than some lost them again! *Phlegethontia* was an amphibian that looked like a snake, having lost its legs entirely. It probably spent much of its time swimming in the water or burrowing in mud where legs would have been more of a nuisance than a help.

FLOURISHING FORESTS

As the amphibians went further inland, they found a landscape of lush forest. This was the Carboniferous age, from 359 to 299 mya. In this time, the first giant trees grew and died. Over millions of years, they turned into the coal we now dig up and burn.

Remaking the world

The first land-plants were small and rootless, and when they died, they were broken down by microbes into the first soil. Soon, a thick enough layer of soil built up to support larger plants with roots. Their stems grew thicker and stronger, supported by tough cells that carried water and nutrients up from the roots. But they did more than make soil. Plants remade the environment and then evolved into it. Their leaves photosynthesized, filling the air with oxygen, just as the cyanobacteria had done millions of years before.

Texture on the bark

Sigillaria
(si-JILL-ah-ree-ah)

Lepidodendron
(LEP-ee-do-DEN-dron)

Giant moss, ferns, and horsetails

Plants grew larger and stronger until around 385 mya they became the first trees. Soon, several types of huge trees towered up to 40 m (130 ft) tall. Unlike modern trees, they were giant versions of smaller plants—clubmosses, ferns, and grass-like plants called horsetails. The clubmoss trees, such as *Lepidodendron*, grew vast. They had a tall stem supported by scaly bark. For half the tree's life it was just a pole. Then, it grew branches covered with narrow leaves arranged in spirals, and with cones carrying seeds.

Prototaxites

Land of giants

Not just the trees grew large. Photosynthesizing plants put more oxygen in the air than there has ever been. Arthropods grew huge, such as the giant dragonfly *Meganeura*, 70 cm (27 in) across. *Pulmonoscorpius* was a scorpion 70 cm (27 in) long, and *Arthropleura*, a myriapod 2.6 m (8.5 ft) long, scuttled through the undergrowth.

Meganeura

Pulmonoscorpius
(PUL-mon-oh-SCOR-pee-us)

EVOLUTIONARY PUZZLES

Prototaxites

One of the strangest things to grow on the newly colonized land was *Prototaxites*. Before the trees, 400 mya, this giant spike of fungus grew up to 8 m (26 ft) tall and 1 m (3 ft) across. It was the largest living thing on land, and one of the strangest organisms of all time.

Medullosa
(med-uh-LOW-sah)

Medullosa

The tree-ferns were like overgrown ferns. *Medullosa* grew to 3.5 m (11.5 ft), growing in the same way as a modern fern by unfurling new fronds at the end of its stem. Unlike modern ferns, it had seeds the size of a hen's egg!

RULING THE LAND

As the amphibians had to live near water, inland areas were at first left to the arthropods. However, other animals soon evolved to move into them—the reptiles.

Becoming reptiles

There's more to living on land than not needing to be wet. Reptiles needed limbs better adapted to moving on land than to skulking in mud and swimming. One adaptation was to swap cartilage for the stronger, stiffer bone. Another was thick, scaly skin. They needed mouths and teeth adapted to catching and eating land-based food, such as arthropods, plants, and smaller reptiles. Most importantly, they needed a way of reproducing without going back to the water.

A change for the better

Like their fishy ancestors, amphibians fertilized their eggs outside the body. The female laid eggs in the water, then the male released sperm over them, fertilizing the eggs, which then developed in the water. For animals to move inland, this had to change. Reptiles developed eggs with a leathery or hard shell that could be laid in the ground without drying out. For this to work, the male had to place the sperm in the egg before it grew its shell. They did this by putting the sperm directly into the female's body. Internal fertilization freed animals from breeding near water.

Varanops (vah-rah-nops), a synapsid.

Dimetrodon (dy-MET-roh-don), a synapsid.

Labidosaurus (lah-bih-doh-SORE-uss), a primitive reptile.

Eryops (EH-ri-ops), an amphibian.

Scutosaurus (SCOOT-ah-SORE-uss), a parareptile.

Strong and stocky

Without the restrictions of the amphibians, reptiles could live anywhere, and did so. But the amphibians didn't disappear. They continued to live in the spaces for which they were best adapted. Large amphibians such as *Eryops* grew to 2 m (6.5 ft) and looked rather like a chunky crocodile.

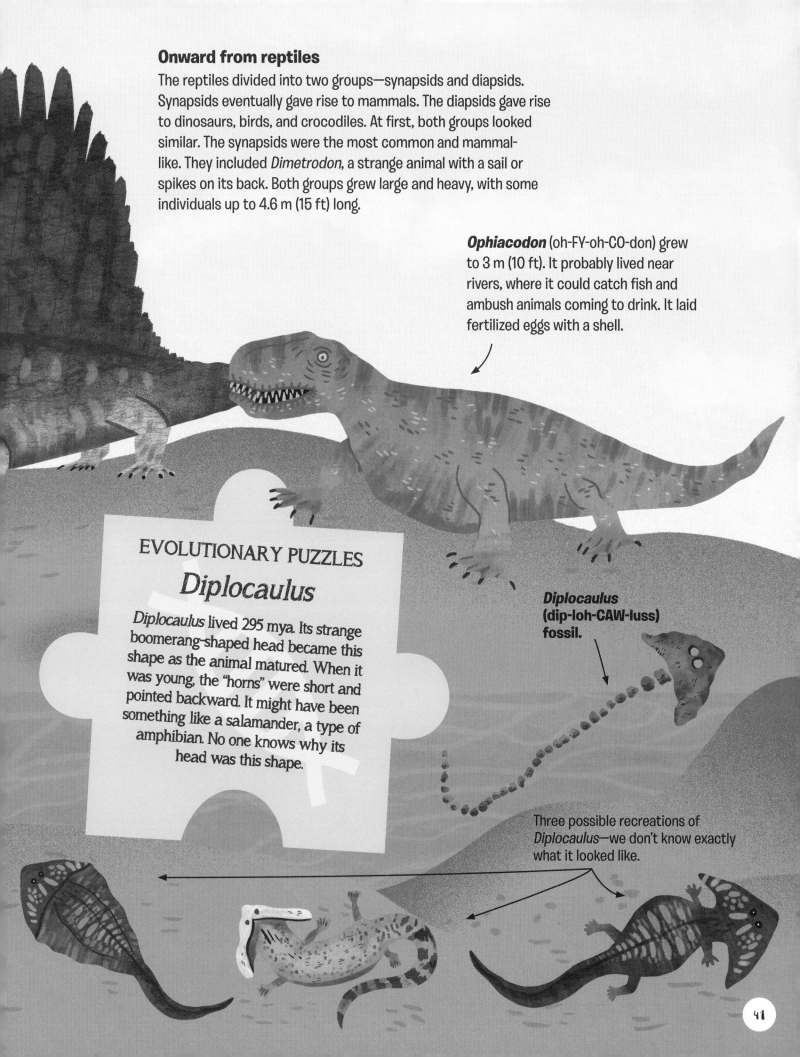

Onward from reptiles

The reptiles divided into two groups—synapsids and diapsids. Synapsids eventually gave rise to mammals. The diapsids gave rise to dinosaurs, birds, and crocodiles. At first, both groups looked similar. The synapsids were the most common and mammal-like. They included *Dimetrodon*, a strange animal with a sail or spikes on its back. Both groups grew large and heavy, with some individuals up to 4.6 m (15 ft) long.

Ophiacodon (oh-FY-oh-CO-don) grew to 3 m (10 ft). It probably lived near rivers, where it could catch fish and ambush animals coming to drink. It laid fertilized eggs with a shell.

EVOLUTIONARY PUZZLES
Diplocaulus

Diplocaulus lived 295 mya. Its strange boomerang-shaped head became this shape as the animal matured. When it was young, the "horns" were short and pointed backward. It might have been something like a salamander, a type of amphibian. No one knows why its head was this shape.

Diplocaulus (dip-loh-CAW-luss) fossil.

Three possible recreations of *Diplocaulus*—we don't know exactly what it looked like.

41

BONE TO STONE

We can work out which plants and animals have lived in the past because we have fossils of them. Fossils are imprints in stone, or parts of organisms that have turned to stone themselves, under just the right conditions.

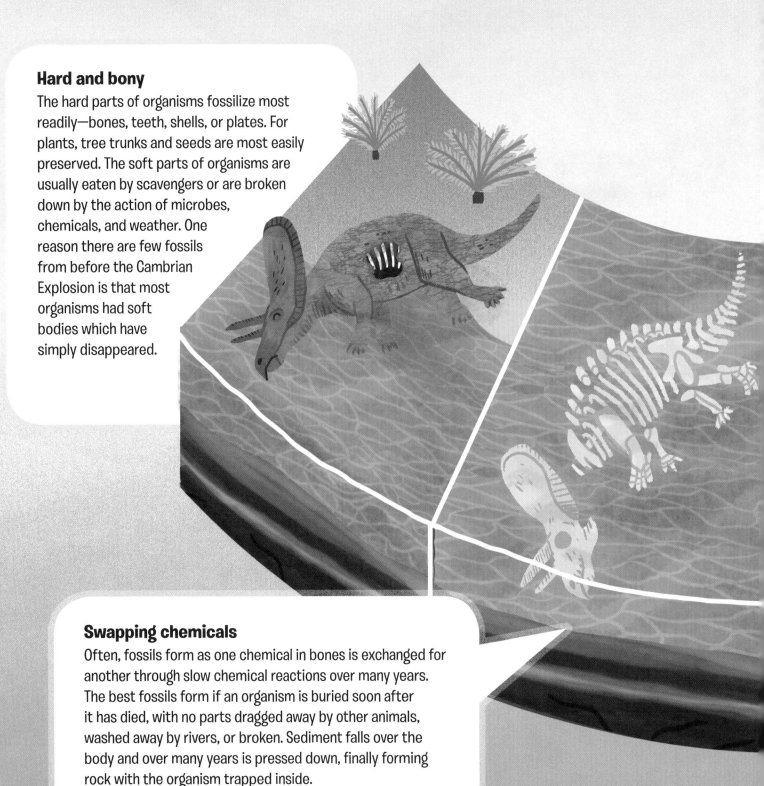

Hard and bony

The hard parts of organisms fossilize most readily—bones, teeth, shells, or plates. For plants, tree trunks and seeds are most easily preserved. The soft parts of organisms are usually eaten by scavengers or are broken down by the action of microbes, chemicals, and weather. One reason there are few fossils from before the Cambrian Explosion is that most organisms had soft bodies which have simply disappeared.

Swapping chemicals

Often, fossils form as one chemical in bones is exchanged for another through slow chemical reactions over many years. The best fossils form if an organism is buried soon after it has died, with no parts dragged away by other animals, washed away by rivers, or broken. Sediment falls over the body and over many years is pressed down, finally forming rock with the organism trapped inside.

Who was here?

Not all fossils are of body parts. Trace fossils are marks left by organisms. They include footprints, tail-drag marks, and holes, burrows, or nests made by animals. Sometimes fossils can give us clues about how animals behaved, and sometimes scientists can make intelligent guesses by looking at how similar modern organisms live, but there is a lot we will never know.

MIND THE GAP

There are some gaps in the fossil record where scientists have found few fossils. There's a gap at just the point when amphibians were developing. When *Tiktaalik* was found in 2004, it was the first fossil of a "fishapod" on the way to living on land. Gaps in the fossil record do not mean there were no organisms at those times, just that we have no fossils of them.

More lost than found

Very few of the organisms that have ever lived become fossils. And very few of the fossils that have ever formed have been found, or will ever be found. For us to find fossils, they must usually be pushed up to near the surface by geological activity—movement of the rocks in Earth's crust. Rocks that are now deep below ground, under cities, or at the bottom of the sea must hide countless fossils we will never see.

DOOM AND GLOOM

Around 252 mya, at the end of the Permian period, the land was covered with forest that was home to giant arthropods and large land animals. The seas were teeming with fish, trilobites, and ammonites, and the swamps sheltered large and small amphibians. But then something went wrong. One of the worst extinction events in Earth's history wiped out 90–95 percent of all plant and animal species.

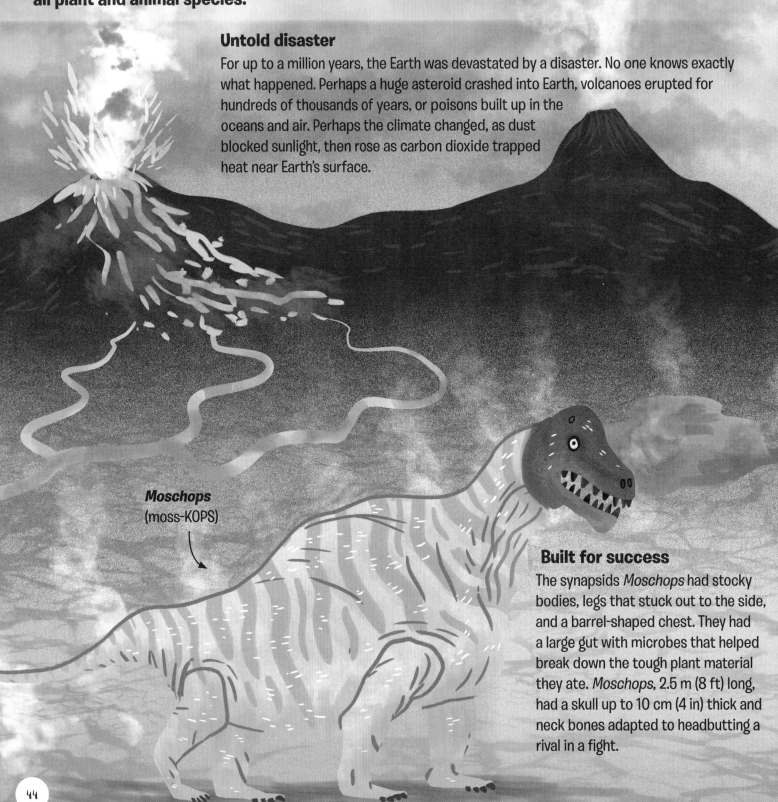

Untold disaster

For up to a million years, the Earth was devastated by a disaster. No one knows exactly what happened. Perhaps a huge asteroid crashed into Earth, volcanoes erupted for hundreds of thousands of years, or poisons built up in the oceans and air. Perhaps the climate changed, as dust blocked sunlight, then rose as carbon dioxide trapped heat near Earth's surface.

Moschops
(moss-KOPS)

Built for success

The synapsids *Moschops* had stocky bodies, legs that stuck out to the side, and a barrel-shaped chest. They had a large gut with microbes that helped break down the tough plant material they ate. *Moschops*, 2.5 m (8 ft) long, had a skull up to 10 cm (4 in) thick and neck bones adapted to headbutting a rival in a fight.

Feast and famine

When the sunlight is blocked and temperatures fall, organisms suffer. Plants can't photosynthesize without light, and they die. That leaves nothing for herbivores to eat, so they die, too. Then the carnivores that eat the herbivores die. Scavengers first feast on the dead, but when the bodies run out, they die, too. (Cheery stuff, right?) Microbes and fungi break down what is left. Large quantities of fossilized fungi suggest they had a feast 252 mya.

THRIVING ON DISASTER

There's always something ready to benefit from a catastrophe. In evolutionary science, they are called "disaster taxa." One of the most important after the end-Permian extinction was *Lystrosaurus*, a synapsid that ate plants. About the size of a sheepdog, it had large tusks, which it probably used for digging out roots, and perhaps for digging a burrow. It is one of the most common fossils from the period following the extinction.

Back from the brink

After this disaster, the few survivors adapted quickly to take advantage of the extra places to live, water, and food. Freedom from predators helped new species to become established. New soil made by decomposers gave surviving seeds somewhere to sprout when the air cleared.

The years following the extinction event saw new species evolving quickly. With most of the large amphibians and reptiles gone, there was space for a new kind of animal—the dinosaurs.

Lystrosaurus
(lye-stroh-SORE-uss)

DINOSAURS AND FRIENDS

Just 5 percent of species survived the extinction event at the end of the Permian period, but in the years after it (the Triassic period) the world filled with new life, on the land, in the sea, and even in the air.

New reptiles evolved, called archosaurs. They were the ancestors of both dinosaurs and the reptiles we have today, such as crocodiles, lizards, and snakes. Reptiles went back into the water and even took to the air. And the mammal-like reptiles? They didn't disappear, but they became much smaller, and lived in the tiny spaces other animals didn't use—burrows underground and the tops of trees. Their day would come, but not just yet. First, the largest animals ever to walk the Earth would rule the land for more than 150 million years.

North America, 76 mya. Clockwise, from top:
Pteranodon (teh-RAH-no-don), **Gryposaurus monumentensis** (GRIP-oh-SORE-uss MAHN-you-men-TEN-siss), **Corythosaurus casuarius** (cor-IH-tho-SORE-uss cas-yoo-AH-ree-us), **Panoplosaurus** (PAN-op-lo-SORE-uss).

BACK INTO THE WATER

Just like the land, the sea lost most of its organisms in the end-Permian extinction. There was plenty of room for new arrivals, and some land animals returned to the water.

Evolution in reverse

Animals had adapted to life on land by changing their bodies. Floppy, cartilaginous fins became stiffened legs with jointed elbows and knees and movable fingers and toes. Bodies grew to a stocky shape for pushing through undergrowth and conserving heat and water. As reptiles went back to the water, they adapted again. Their legs became flipper-like, and their bodies and heads grew streamlined.

Plesiosaurus
(PLEH-see-oh-SORE-uss)

Cymbospondylus
(SIM-boh-SPON-dih-luss)

LIVE BABIES

All amphibians and most land-going reptiles laid eggs. Marine reptiles gave birth to live young. Live birth probably evolved more than 100 times independently in different species of reptiles.

Like a fish—but not

The most fish-like marine reptiles were ichthyosaurs. *Cymbospondylus* was an early ichthyosaur. Its limbs looked like paddles, but it was an evolutionary work-in-progress. The later ichthyosaurs had a fin on the back (dorsal fin) that helped them to cut through the water, and a tail fluke that helped them to move along. They looked rather like modern dolphins. Like dolphins, they had to come to the surface to breathe air. They grew bigger and bigger, until 205 mya, when some were almost as big as a blue whale.

Ichthyosaurus
(ICK-thee-oh-SORE-uss)

Long and short necks

Other shapes work in the water besides fish-shapes. The plesiosaurs and pliosaurs looked more like water-going dinosaurs than fish, yet they were streamlined. Plesiosaurs had long necks—extremely long in the case of *Elasmosaurus*—and a small head. The fierce pliosaurs like *Liopleurodon* had a large head and short neck, more like a crocodile.

Elasmosaurus
(el-LAZZ-moh-SORE-uss)

Liopleurodon
(LEE-oh-PLUR-oh-don)

Mosasaurus
(MOH-sah-SORE-uss)

Reinventing the reptile

A new type of sea reptile, the mosasaurs, appeared about 100 mya and took over as the others died off. They evolved from a new set of land-going reptiles returning to the sea. Sometimes called the *T. rex* of the sea, mosasaurs are among the most dangerous animals that have ever lived in the oceans. Smaller ones ate squid and fish, but the largest ate diving seabirds, sharks, and other mosasaurs. Their jaw could unhinge, like that of a snake, allowing them to gulp down large prey whole.

STRANGELY SIMILAR

The marine reptiles grew streamlined bodies, fins, and tail flukes. These make moving in water easy, and fish have them, too. Organisms often develop in similar ways if they face the same problems and challenges. It's called convergent evolution.

Just keep swimming

Fish evolved from ancestors that were torpedo-shaped or ribbon-like. They had nothing that made it hard to move through water—no legs, sticky-out ears, or even fins at first. When fish evolved, they kept the streamlined shapes but added fins and flukes for control and speed.

Common seal

Pliosaurus
(PLEE-oh-SORE-uss)

Kronosaurus
(CRON-oh-SORE-uss)

Water bodies

When later animals moved back from land to sea, their needs changed. In air, an animal needs a body that holds its shape with strong bones and muscles, while in the sea, the water holds the animal up. Animals can move through air easily, but moving through water takes more effort; a streamlined shape helps. Land animals have adapted to life in the water by changing shape, turning limbs into flippers, and losing fur or feathers to move smoothly. Marine reptiles, marine mammals such as a dolphins and seals, and diving birds such as penguins, have all evolved similar solutions separately.

Bottle-nosed dolphin

Flying by air and sea

Many different kinds of animals fly—birds, bats, insects, pterosaurs (extinct flying reptiles), and even "flying" fish. They have all developed flight separately, another example of convergent evolution. They don't all fly in the same way. Some animals glide, spreading out a web of skin and parachuting. Birds, insects, and bats flap their wings. Eagles soar high above land, riding the air currents. Some animals "fly" underwater. A ray moves by slowly flapping its "wings" up and down.

Blue morpho butterflies

Hummingbird

Penguin

Ray

BEHAVE YOURSELVES!

Convergent evolution also happens in the way in which creatures behave. Some fish swim in huge shoals that keep their formation and change direction smoothly. In the sky, birds do the same, collecting in flocks that swoop and dive together, often in shapes like fish shoals.

Shark

RISE OF THE DINOSAURS

While some reptiles returned to the seas, others stayed on land, quickly becoming more and more varied.

Divide and conquer

The surviving reptiles, such as *Lystrosaurus*, spread out across the world. They could do this easily, as all the land was in a single, vast continent called Pangaea. Within just 5 million years, the archosaurs—or "ruling reptiles"—had evolved. One group would become the dinosaurs and birds; the other would become crocodilians—reptiles that look like crocodiles.

Lystrosaurus

Proto-birds and proto-crocodiles

The bird-like group were slim and speedy. The earliest dinosaurs were in this group. They included one of the first, *Herrerasaurus*, from 228 mya. Soon after came fast-running predators like *Coelophysis* and plant-eating dinosaurs like *Eocursor*. The plant-eaters became the ornithischians—dinosaurs with beak-like mouths that eventually came in lots of shapes and sizes. A third group, the sauropodomorphs (which just means "shaped like a sauropod"), were heavy and slow. Up to 10 m (33 ft) long, *Riojasaurus* and *Plateosaurus* walked on four legs, like the later giant Jurassic sauropod *Diplodocus*.

Riojasaurus (ree-OH-ha-SORE-uss)

Herrerasaurus (her-AIR-ah-SORE-uss)

Coelophysis (SEE-low-fy-siss)

Eocursor (EE-oh-cur-suh)

Plateosaurus
(PLAT-ee-oh-SORE-uss)

Ankles for archosaurs

The bird-like archosaurs had a hinged ankle that changed the shape of the foot and let them run quickly. The crocodilian ankle is like a ball that fits into a socket. It has a wider range of movement, but isn't good for speed. That's one of the reasons why a crocodile can't run like an ostrich.

Lookalikes

Most crocodilians were stocky and slow meat-eaters. Chunky animals like *Postosuchus*, about 4 m (13 ft) long, would have been formidable threats to the plant-eaters. With its narrow snout and powerful jaws, it looked quite like *T. rex*. It lived in the same way and had the same role in its world. This is an example of convergent evolution, as the much-later *T. rex* came from the other bird-like group.

Another crocodilian grew more like the thin, nimble early theropods. *Effigia*, with its long neck and tail, narrow head, and snippy mouth, looked like *Coelophysis* but had a toothless beak and ate mostly plants.

Effigia (eff-ih-GEE-ah)

Postosuchus (poss-toh-SOOK-uss)

Dino-shapes

Dinosaurs came in three basic designs:

Theropods were hunters. They had longer back legs than front legs and usually ran on their back legs.

Sauropods ate plants. They usually walked on all four legs and many grew to huge sizes.

Ornithischians were all the rest, including stegosaurs, ankylosaurs, and dinosaurs like *Triceratops* and *Parasaurolophus*. They ate plants.

AGE OF GIANTS

The dinosaurs started off a decent size, but grew and grew. By the middle of the Jurassic (200–145 mya), the sauropod dinosaurs were the largest animals ever to live on land.

Why be big?

Sauropods grew enormous several times in different parts of the world, so it must have helped them to survive. Several features made it possible. Their bones and bodies had air-spaces, which made them light for their size, so they could grow bigger without becoming too heavy to support their own weight. Their long necks meant they could reach food that was out of reach of other animals, and they had a small head which a long neck could support easily. They stripped leaves and twigs and swallowed them without chewing, so they could eat a lot quickly. Being bigger became an advantage—they could reach even more food, and were too large for predators to attack.

Big, bigger, biggest

Sauropods started quite small—*Riojasaurus* was just 10 m (33 ft) long. But that's still the size of a bus. As time passed, they grew larger. *Camarasaurus* grew to 15 m (49 ft), and *Apatosaurus* to 22 m (72 ft). *Diplodocus* could be 25 m (82 ft) long, but the giant *Argentinosaurus* grew to 37 m (121 ft). It might not even have been the biggest!

Apatosaurus
(ah-PAT-oh-SORE-uss)

Camarasaurus
(KAM-uh-rah-SORE-uss)

Magyarosaurus
(MAG-yah-ro-SORE-uss)

Four legs good, two legs (sometimes) bad

A large body needs a lot of support. The large sauropods kept all four feet firmly on the ground. Some of the smaller types might have stood on two legs occasionally to reach higher in the trees, but the biggest couldn't (and didn't need to). Other large, heavy dinosaurs, such as *Triceratops* and *Ankylosaurus*, also kept all their feet on the ground. *T. rex*, though large, was much smaller than the sauropods. It had huge rear legs to support all its weight on two legs. All theropods walked on two legs.

Snacking sauropods

If sauropods had all eaten the same things, there would not have been enough food to support lots of them. Instead, different species ate different plants and had different mouths and teeth suited to the foods they ate. Some had peg-shaped teeth, which they used as a comb to strip leaves from branches. Others had broader teeth with sharp edges for cropping through branches.

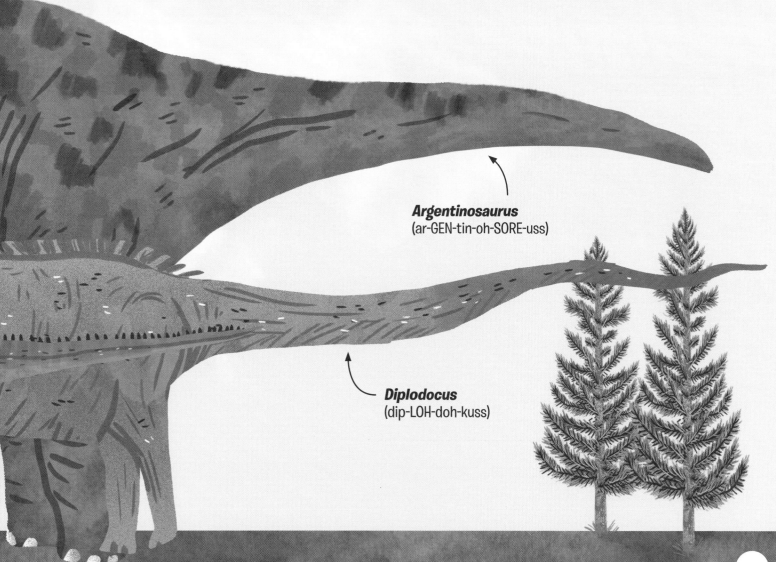

Argentinosaurus
(ar-GEN-tin-oh-SORE-uss)

Diplodocus
(dip-LOH-doh-kuss)

SHIFTING LANDS

The land doesn't stay in one place. Slowly, it drifts, breaking apart and regrouping over millions of years. At the end of the Permian, all the land was clumped into a single continent, Pangaea. That meant organisms could spread all over the world.

Floating lands

Earth's surface is a solid, rocky crust floating on a very thick layer of hot, semi-molten rock called magma. The crust is broken into chunks, called tectonic plates. These fit together like the pieces of a jigsaw puzzle. The magma moves around the Earth, carrying the plates of crust with it. However, because of currents in the magma they don't all move together. Over hundreds of millions of years the land rearranges itself, sometimes forming a single large continent surrounded by a global ocean and sometimes, as now, broken into separate continents surrounded by linked oceans.

Growing up alone

When the land is all grouped together, plants and animals can spread all over it. But when it is broken apart, most cannot cross large areas of sea. Today, only birds, insects, or small animals carried on the wind or floating on objects in the sea could reach Australia from elsewhere. When areas of land break apart, the organisms on separate chunks develop separately.

On islands, the effect is especially noticeable. Many island species evolve into smaller versions of mainland organisms. Sauropods that lived on islands were often quite small. The island sauropod *Magyarosaurus* (MAG-yah-ro-SORE-uss) was just 6 m (20 ft) long, while the giant sauropods were five times as long.

SPIKES AND SUCHLIKE

Stegosaurus lived 163–100 mya on land that is now in the western USA. But other stegosaurs have been found in China, Africa, and Europe. The earliest stegosaurs could walk across the land to spread out. When Pangaea broke up, starting about 175 mya, the stegosaurs stranded on different continents evolved separately, adapting to conditions in their own areas.

Kentrosaurus
(KEN-tro-SORE-uss)

Stegosaurus
(STEG-oh-SORE-uss)

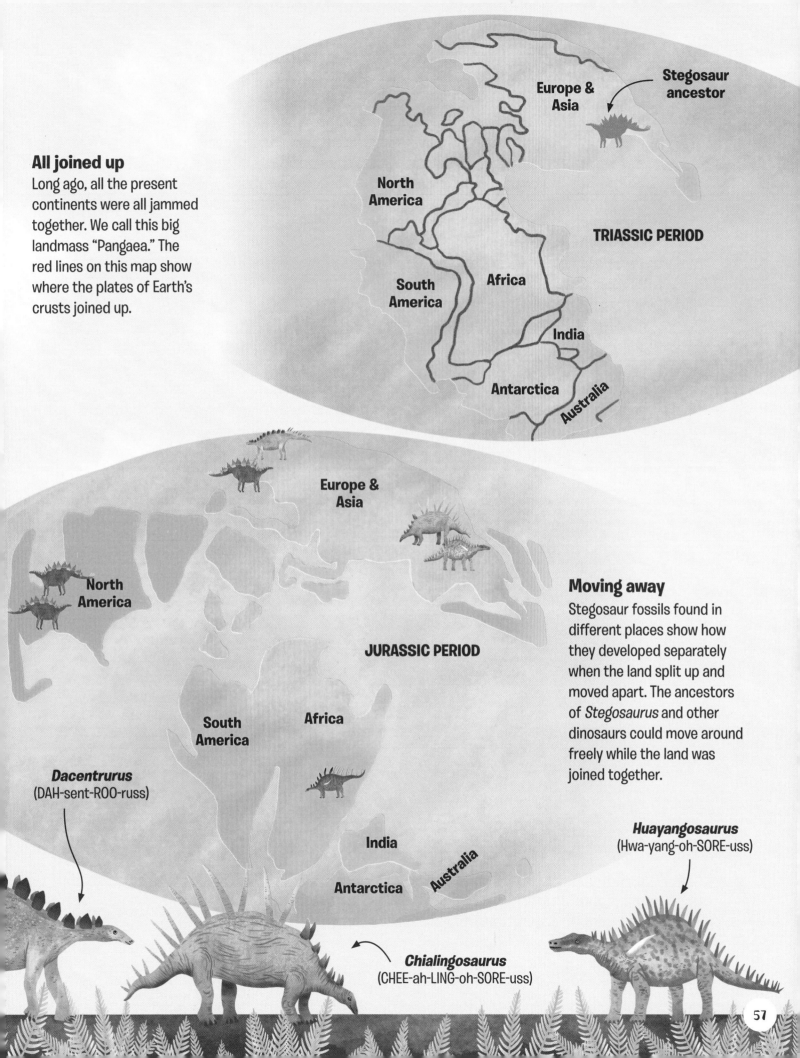

All joined up

Long ago, all the present continents were all jammed together. We call this big landmass "Pangaea." The red lines on this map show where the plates of Earth's crusts joined up.

Europe & Asia

Stegosaur ancestor

North America

TRIASSIC PERIOD

South America

Africa

India

Antarctica

Australia

Europe & Asia

North America

JURASSIC PERIOD

South America

Africa

India

Antarctica

Australia

Moving away

Stegosaur fossils found in different places show how they developed separately when the land split up and moved apart. The ancestors of *Stegosaurus* and other dinosaurs could move around freely while the land was joined together.

Dacentrurus
(DAH-sent-ROO-russ)

Chialingosaurus
(CHEE-ah-LING-oh-SORE-uss)

Huayangosaurus
(Hwa-yang-oh-SORE-uss)

UP, UP, AND AWAY

While some reptiles returned to the water, others took to the sky. The pterosaurs were flying archosaurs that lived at the same time as the dinosaurs, from around 220 to 66 mya. Though they look something like birds, birds are not descended from them.

Getting off the ground

The pterosaurs evolved from small, agile reptiles. They were the first animals after insects to develop flight. Their wings were flaps of skin stretched between their front and rear legs. The fourth finger grew extra long to support the wing, and the others stuck out at the wing's joint. On the ground, pterosaurs walked on all four limbs with their wings folded.

Many pterosaurs lived at the coast and ate fish. The early *Dimorphodon* had a large head and beak like a modern puffin. It possibly scooped fish from the surface of the sea, as puffins do now. Slightly larger, *Rhamphorhynchus* had scary-looking snaggle teeth, which held fish securely in its narrow beak.

Rhamphorhynchus
(ram-for-INK-uss)

Pterodactylus
(TEH-ro-DACK-tih-lus)

Dimorphodon
(dy-MORF-oh-don)

Anhanguera
(an-HAN-gweh-rah)

58

Heads and tails

The pterosaurs were successful, and there were many different types over 150 million years. Some, like *Tupandactylus* and *Tupuxuara*, lived over land eating the fruit of cycads.

The early pterosaurs had long tails, but most later species lost their tails. *Pterodactylus* (TEH-ro-DACK-tih-lus) had a short tail, longer neck and a small head-crest.

From bird-size to plane-size

The pterosaurs grew bigger, until the largest had the wingspan of a small plane. The giant *Pteranodon* lived like a modern albatross. With a wingspan of 6 m (20 ft), it rarely landed and was usually at sea catching fish in a toothless beak. Flocks of *Pteranodon* probably glided over the shallow seas. The largest was a creature of nightmares. At 12 m (39 ft) across, *Quetzalcoatlus* (KWETS-ul-koh-AT-lus) walked the land eating dinosaurs. It was the largest flying creature ever. Just its head was larger than a tall man.

Pteranodon
(teh-RAH-no-don)

EVOLUTIONARY PUZZLES
Pterosaur wings

The details of pterosaur wings are a mystery. They might have stretched to the ankles of the rear legs, or just to the knees, or even the hips. Bats have evolved similar wings, which attach to the ankle of the rear leg and often to the tail

Tupuxuara
(too-poo-CHOOR-ah)

Heading for trouble

Head crests grew larger and more ornate. Some species had huge crests that could have been a nuisance. The giant *Tupandactylus*, with a wingspan of 5.5 m (18 ft), might have struggled with its crest.

Tupandactylus
(too-pan-DACK-tih-lus)

SMALL BEGINNINGS

Alongside the giant dinosaurs, much smaller creatures quietly carved out a space for themselves. These became the mammals, of which humans are a late example. They began as small, rodent-like animals that lived in the trees or underground.

Fast and furry

Mammals evolved from a group of mammal-like reptiles called cynodonts. Over time, they developed the distinctive features of mammals—being warm-blooded, having fur or hair, and producing milk for their young. Warm-blooded animals can control their body temperature—they don't need the sun to warm them. This meant early mammals could be active in the cool night, when predators were asleep.

The first mammals ate insects and small lizards, animals too small for most dinosaurs, and were the first animals besides arthropods to live in the treetops.

The size of a large dog, *Repenomamus* (reh-pen-oh-MAM-uss) lived 125 mya in China. It ate tiny dinosaurs.

Specialists

Animals that are successful find a niche for themselves—their own space in the natural world, with somewhere to live, something to eat, and a lifestyle that suits their surroundings. Early mammals evolved in different ways to fill available niches. Many had features that evolved independently in much later animals.

Tiny *Morganucodon* (mor-GAN-uh-COH-don) was not quite a mammal, as it had a type of jaw found in reptiles. It ate insects, probably had hair, and laid eggs with a leathery shell. It lived 205 mya.

Sinoconodon (SY-no-CON-oh-don) was also nearly a mammal, larger than usual at 30 cm (1 ft).

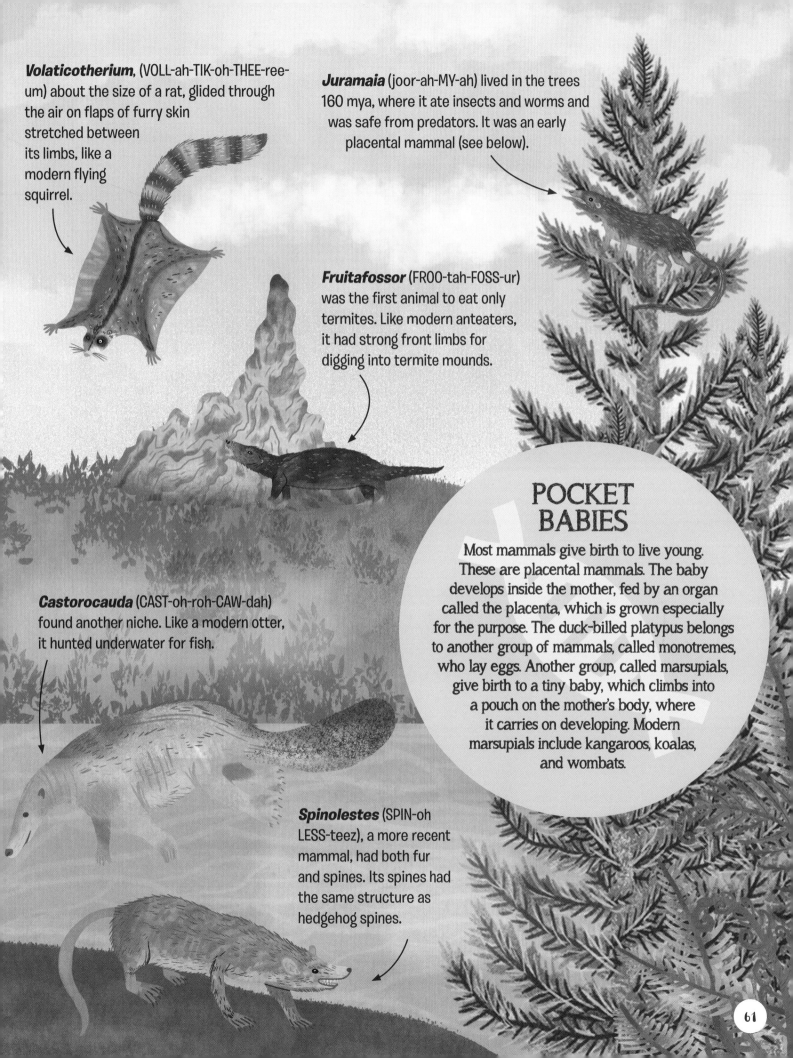

Volaticotherium, (VOLL-ah-TIK-oh-THEE-ree-um) about the size of a rat, glided through the air on flaps of furry skin stretched between its limbs, like a modern flying squirrel.

Juramaia (joor-ah-MY-ah) lived in the trees 160 mya, where it ate insects and worms and was safe from predators. It was an early placental mammal (see below).

Fruitafossor (FROO-tah-FOSS-ur) was the first animal to eat only termites. Like modern anteaters, it had strong front limbs for digging into termite mounds.

Castorocauda (CAST-oh-roh-CAW-dah) found another niche. Like a modern otter, it hunted underwater for fish.

POCKET BABIES

Most mammals give birth to live young. These are placental mammals. The baby develops inside the mother, fed by an organ called the placenta, which is grown especially for the purpose. The duck-billed platypus belongs to another group of mammals, called monotremes, who lay eggs. Another group, called marsupials, give birth to a tiny baby, which climbs into a pouch on the mother's body, where it carries on developing. Modern marsupials include kangaroos, koalas, and wombats.

Spinolestes (SPIN-oh LESS-teez), a more recent mammal, had both fur and spines. Its spines had the same structure as hedgehog spines.

DINOSAURS TAKE FLIGHT

Pterosaurs were the first vertebrates to fly, but they soon had to share the airspace with birds. Birds evolved from theropod dinosaurs; they are the only dinosaurs still living.

Feathers without flight

Lots of theropod dinosaurs were feathery. *T. rex* might have had fluff as a baby, even if it didn't have a full set of feathers when grown up.

Originally, feathers helped dinosaurs to keep warm. This was especially useful for small dinosaurs that lost their heat quickly, like *Anchiornis*, just 1 m (3 ft) long. If these dinosaurs were cold-blooded, holding on to body heat would have been important, as cold-blooded animals are only active when warm. Keeping warm would mean they could hunt for longer as the air cooled.

Completely covered with feathers, *Anchiornis* looked as if it had four wings as its back legs had long feathers. But it probably ran over the ground rather than flying, as the front wings were not a good shape for flight. It might have been able to glide, or flap its wings to help speed it along while running.

Archaeopteryx
(ARK-ee-op-teh-RIKS)

Sinornithosaurus
(SY-nor-nih-tho-SORE-uss)

Anchiornis (AN-key-OR-niss)

Teeth and tails

Although a feathery dinosaur looks quite like a bird, there are important differences. Even *Archaeopteryx*, part-way between dinosaur and bird, was very different from modern birds. *Archaeopteryx* had bones in its tail, while modern birds have only tail feathers, all attached to the end of the body. Modern birds have toothless beaks, but *Archaeopteryx* had teeth in its jaws. And *Archaeopteryx* had fingers with claws on its wings. It could use these to grasp prey or hold onto a tree.

Scarlet macaw

Confuciusornis (kon-FYOO-shuss-ORN-iss) was the first bird to have a beak.

BIG BIRDS DON'T FLY

The early birds were all small, as far as we know. Later, some grew much larger, but probably didn't fly. We have large birds now that don't fly, including ostriches and emus. These look very like the dinosaurs they have descended from.

Becoming a bird

For birds to become true birds, they had to lose their tail bones and teeth, develop proper beaks, lose the claws on their wings, and turn one of their toes around. A bird has one toe facing backward and opposable, which makes it possible for the bird to perch on a branch.

TOWARD THE MODERN WORLD

All things must come to an end, even dinosaurs. The end came for most dinosaurs 66 mya, when a huge asteroid or comet smashed into the sea near Mexico. Like previous extinction events, the catastrophe cleared the way for other organisms to seize their chance.

The mammals crept out of their burrows and nests to occupy the spaces freed by the dinosaurs and other reptiles. At first they were all small, but soon larger mammals evolved. They spread around the world and, like the reptiles before them, even returned to the sea, becoming the ancestors of whales and dolphins.

Around 5 million years after the death of the dinosaurs, another huge upheaval took place. The climate first became hotter and then cooler, until about 34 million years ago Antarctica froze over. The continents drifted toward their current positions, pushing up great mountain ranges in America and Asia. Those mountains changed the movement of air, leading to more rain, and that changed conditions inland. The modern world—its landmasses, climate, and organisms—was made in the last 60 million years.

DEATH TO ALL DINOSAURS!

The largest dinosaurs lived in the Jurassic, but the Cretaceous (145 to 66 mya) saw some of the most famous, including *Tyrannosaurus rex*, *Triceratops*, *Parasaurolophus*, and *Ankylosaurus*. The dinosaurs were at their most diverse and varied—but it wouldn't last.

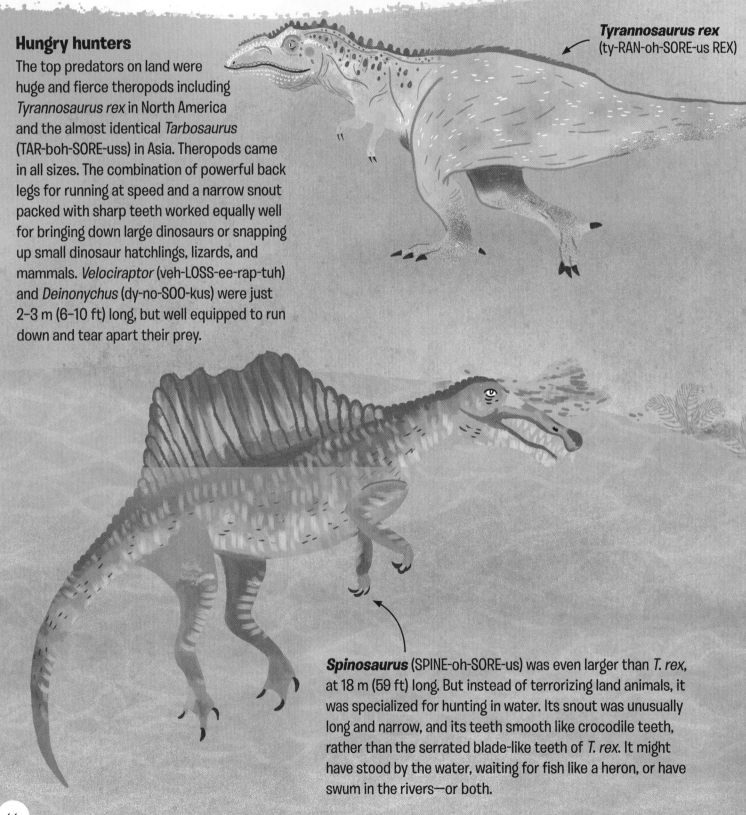

Hungry hunters

The top predators on land were huge and fierce theropods including *Tyrannosaurus rex* in North America and the almost identical *Tarbosaurus* (TAR-boh-SORE-uss) in Asia. Theropods came in all sizes. The combination of powerful back legs for running at speed and a narrow snout packed with sharp teeth worked equally well for bringing down large dinosaurs or snapping up small dinosaur hatchlings, lizards, and mammals. *Velociraptor* (veh-LOSS-ee-rap-tuh) and *Deinonychus* (dy-no-SOO-kus) were just 2–3 m (6–10 ft) long, but well equipped to run down and tear apart their prey.

Tyrannosaurus rex
(ty-RAN-oh-SORE-us REX)

Spinosaurus (SPINE-oh-SORE-us) was even larger than *T. rex*, at 18 m (59 ft) long. But instead of terrorizing land animals, it was specialized for hunting in water. Its snout was unusually long and narrow, and its teeth smooth like crocodile teeth, rather than the serrated blade-like teeth of *T. rex*. It might have stood by the water, waiting for fish like a heron, or have swum in the rivers—or both.

Triceratops
(try-SEH-ra-tops)

Staying alive

Plant-eating dinosaurs evolved different ways to avoid becoming a meal for the fearsome predators. *Triceratops*, heavy and stocky, had long, sharp horns that could pierce the toughest dinosaur hide.

Parasaurolophus (PA-ra-sore-OL-off-uss) could run very fast—possibly faster than *T. rex*—and lived in herds, which reduced the risk for any one individual of becoming a snack for a hungry theropod.

Ankylosaurus (an-KIH-loh-SORE-us) was covered with hard knobs and plates and had a club-like bony tail that it could thrash from side to side, swiping any predator that came too close.

EVOLUTIONARY PUZZLES
Therizinosaurus

Therizinosaurus looks scary, with its long claws like curved swords, but it was probably a plant eater. No one knows what it used these claws for. Perhaps they were for pulling branches toward it so it could munch the leaves, or to defend itself against predators.

Therizinosaurus
(THEH-rih-ZIN-oh-sore-us)

Time's up

An underwater crater near the coast of Mexico is all that's left of the asteroid or comet that struck the Earth 66 mya. The dinosaurs were not the only ones to suffer. But not everything died. Birds, turtles, small mammals, arthropods, many large sea creatures, and a lot of plants survived. They would take over the world.

AFTER THE DINOSAURS

To survive, organisms had to live through both the immediate danger and later effects of the disaster. That could have meant acid rain, wildfires, and dust clogging the air. At first, it would have been cold as the sun was blotted out by dust and smoke, but then it grew hotter as carbon dioxide trapped heat near Earth's surface.

Hot days in the forest

The world grew hotter steadily (and sometimes dramatically) between 60 and 50 mya. The seas were warm, and the poles free of ice. The sea level rose and it rained—a lot. Tropical rain forest covered most of the world. Plants and animals adapted to hot, humid conditions flourished.

CHANGING CLIMATE

The heat and plentiful water were good for plants. One result was that plants such as *Azolla* or mosquito fern, spread widely across the ice-free Arctic. They took in carbon dioxide, and when the plants died and formed sediment, the carbon they had absorbed was locked away. The carbon dioxide in the air reduced and the temperature fell. By 35 mya, there was ice at the South Pole again.

Diacodexis (DY-ah-coh-DEX-iss) was about the size of a modern squirrel and lived in the undergrowth. It could probably run and jump like a deer, but with a long tail to help it balance. It's the first even-toed ungulate—an animal like pigs or sheep which walks on its third and fourth toes.

Azolla (ah-ZOH-lah), also called the mosquito fern.

Plesiadapis (PLEH-see-ah-dap-iss) looked like a lemur and lived in the trees. It is one of the earliest primates—the group of animals that includes humans, gorillas, and monkeys.

Pakicetus (PAH-kih-SEE-tuss) was the size of a wolf. It lived in the forests and shoreline of what is now Pakistan, where it ate meat and fish.

The giant bird **Gastornis** (gas-TORN-iss) stood nearly 2 m (nearly 7 ft) tall. Unable to fly, it stalked the undergrowth, possibly hunting small mammals or eating fruit and large nuts which it cracked with its powerful beak. When the forest gave way to grasslands, *Gastornis* died out. It couldn't run fast enough to escape large predators on an open plain.

Presbyornis (PRESS-bee-OR-niss) was a water bird, as tall as a modern heron at 1 m (3 ft). It's an ancestor of modern ducks, but had much longer neck and legs than a duck.

The crocodilians survived the extinction event well. Animals like **Pristichampsus**, (priss-tee-CHAMP-suss) 3 m (10 ft) long, stalked the waterways, snatching animals and birds that came to drink there.

TAKING THE PLUNGE

As the mammals grew larger and more varied, some returned to the sea. These evolved into the cetaceans—the whales and dolphins that breathe air while living all the time in water.

Pakicetus

Not very like a whale

Pakicetus patrolled the riverbanks of what is now Pakistan, feeding on meat and sometimes fish. It looked like a regular land-going mammal, with fur, four legs, a long tail, and a long snout. But all these features would be adapted as its descendants returned to the water. *Pakicetus* is the earliest ancestor of the whales.

Ambulocetus still spent a lot of time on land, but it had shorter front limbs than *Pakicetus*, and webbed toes and fingers that helped it to swim. These adaptations made it harder to run on land. As it became better suited to the water and worse suited to life on land, it spent more time in the water. Then it became even better suited to life in the water—and so on.

Ambulocetus
(AMB-yoo-loh-SEE-tuss)

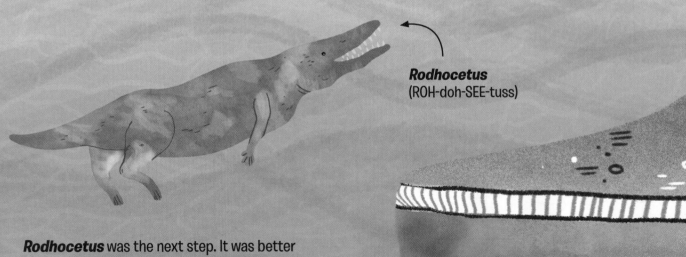

Rodhocetus
(ROH-doh-SEE-tuss)

Rodhocetus was the next step. It was better adapted to swimming than to life on land. Its large body was supported in water and its limbs did not need to be as long—but it did have large, webbed feet. *Rodhocetus* still had fur rather than blubber (fat) to keep it warm and probably swam near the surface.

70

Dorudon (doh-roo-DON) was much more whale-like. At 5 m (16 ft) long, it was sleek and streamlined. Its limbs had shrunk to tiny flippers, and it had a rudder-like tail more suitable for steering underwater than balancing on land. It fed on fish near the bottom of the sea, and—as the oceans are all linked—spread around the world.

Atlantic bottlenose dolphin

Dorudon (doh-roo-DON)

Dorudon had nostrils halfway between the end of its snout and the top of its head. The nostrils of modern whales and dolphins have fused into a single blow-hole above the head. They use echolocation, finding their way in the dark deep sea by making noises and listening to the echoes. *Dorudon* could not do that.

ADAPTATION IN ACTION
From teeth to baleen

The largest modern whales have small or no teeth. They feed by filtering water through a flexible, comb-like veil of baleen. It means whales can feed on the vast supplies of tiny organisms at the bottom of the food chain, eating large quantities at once while using little energy to catch them. It's a trick that helped them to grow ever bigger.

Blue whale

BROWSERS AND GRAZERS

As the world cooled, tropical forest gave way to a mix of woodland and grassland. Grasses first grew under dinosaurs' feet, but the climate was not right for it to spread far. Around 30 mya, tropical forest had shrunk to just a band near the equator, and grasslands took over.

Feeding on the floor

The spread of grasses was good for animals that ate at ground level—and they in turn were good for the spread of grass. Unlike other low-growing plants, grass survives well if it's cropped regularly by grazing animals (or lawnmowers!). It spreads from the root, and cropping the top encourages it to grow more. Plants that spread only by making seed can't reproduce if their seed-bearing parts are cut off.

It might seem odd that a plant benefits from being eaten. However, while the grass just got a haircut when eaten by grazers, the low-growing plants it competed with were killed.

High-speed herds

The first grazing animals were small, like early horses *Mesohippus* in North America and *Eurohippus* in Europe. At just 60 cm (2 ft) tall, they would have been easy prey for dog-like predators that were appearing, such as *Hyaenodon*. On grassland, with nowhere to hide, grazers had to evolve to survive. They grew longer legs, which helped them to run fast, and they often lived in herds. Herding cuts the chance of an animal being taken by a predator—especially if it hides in the middle of the herd. Predators killed the slowest, weakest animals, leaving the stronger, faster ones to breed and pass on the genes for being tall, strong, and fast.

Mesohippus
(MESS-oh-HIP-uss)

Eurohippus
(YOO-roh-hip-uss)

Hyaenodon
(hy-EE-noh-don)

Paraceratherium
(PA-ra-SEE-thee-ree-em)

Big bruisers

Browsers feed from bushes and trees. In the hot forests, browsers grew larger, including the largest land mammal ever, *Paraceratherium*. At 5.5 m (18 ft) tall and 8 m (26 ft) long, it fed from the woodland trees in Asia. The stocky *Megacerops*, like a rhino with a Y-shaped horn, fed on lower-growing vegetation in North America. Each evolved body and teeth adapted to where it fed. But the huge browsers were on the way out as the forests shrank and grasslands spread.

Megacerops
(MEG-ah-SEE-rops)

ADAPTATION IN ACTION
Goodbye to toes

The horses were some of the first ungulates—hoofed animals. Hooves evolved as toes reduced. As grasslands spread, animals could run faster with a hoof over dry, open ground than with a regular, toed foot. Their legs grew longer and their toes eventually fused into a single hoof, giving them the speed they needed to escape predators. In areas with soft, wet, ground, horses tended to have separate toes.

BETTER TOGETHER

Grazers and grasses help each other in a relationship biologists call mutualism. Some organisms evolve together with others—they co-evolve—and each benefits from the other. Flowering plants co-evolved with the insects that pollinate them.

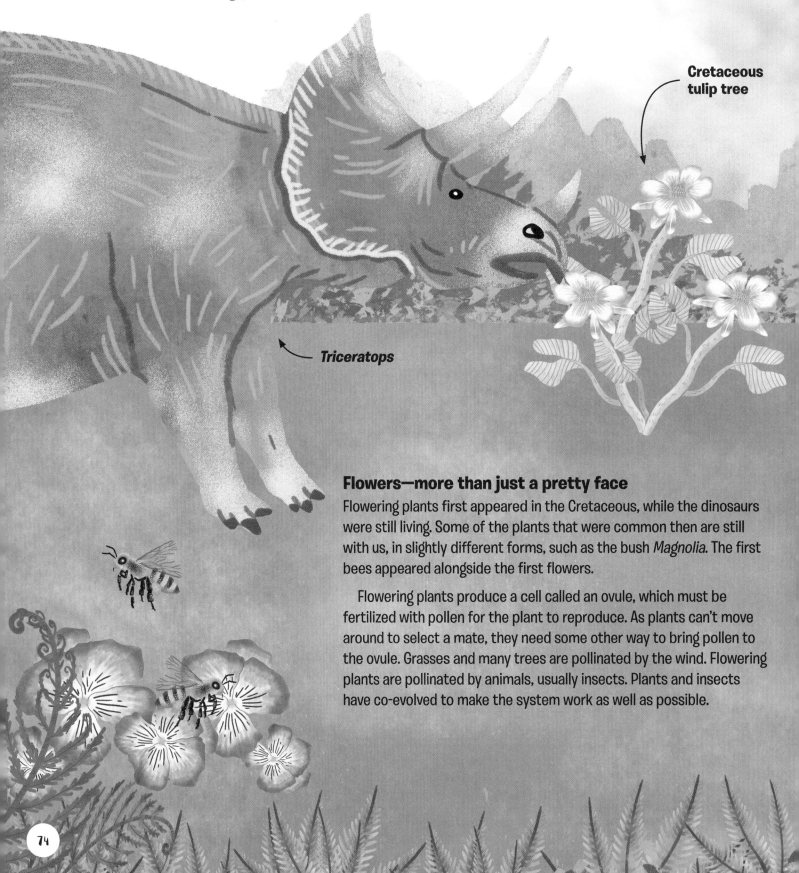

Cretaceous tulip tree

Triceratops

Flowers—more than just a pretty face

Flowering plants first appeared in the Cretaceous, while the dinosaurs were still living. Some of the plants that were common then are still with us, in slightly different forms, such as the bush *Magnolia*. The first bees appeared alongside the first flowers.

Flowering plants produce a cell called an ovule, which must be fertilized with pollen for the plant to reproduce. As plants can't move around to select a mate, they need some other way to bring pollen to the ovule. Grasses and many trees are pollinated by the wind. Flowering plants are pollinated by animals, usually insects. Plants and insects have co-evolved to make the system work as well as possible.

Sowing seeds

A fertilized flower sets seeds, from which new plants can grow. Plants have different ways of dispersing seeds. Some seeds blow in the wind, but some plants rely on animals. Many seeds grow inside fruit, which animals eat. The seed passes straight through the animal's gut and is deposited in droppings, which helps the seed grow. Other plants have sticky or spiky seeds that stick to animals or birds and so hitch a ride.

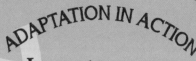

Wood thrush

Bees have evolved "pollen baskets" on their legs. These carry pollen to their hive, where it is used for food. The bees collect the pollen from lots of different individual plants, and enough pollen brushes off into the flowers that they fertilize the plants that they visit.

ADAPTATION IN ACTION
Long tongues for long tubes

Bees and flowers sometimes co-evolve specific paired features. For example, some bees have an extra-long tongue to sip from flowers with nectar kept at the bottom of unusually long tubes. As the tube and tongue lengthen, other insects can't compete with the bee for the nectar.

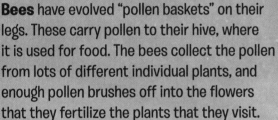

Where the bee sucks

Flowers produce nectar, a sweet syrup, and are often bright shades to attract pollinators. Many have "nectar guides"—lines that point insects to the nectar they eat. On the way to the nectar, insects brush past the stamens that hold pollen, and it sticks to their bodies. Pollen brought from other plants falls off, pollinating the flower.

Some plants are pollinated by birds or bats. They are specially adapted, with larger supplies of nectar and bigger flowers. Bats fly by night, and the flowers that rely on them often open or release scent at night.

PRIMATES IN THEIR PRIME

Humans belong to a group of mammals called primates, which includes monkeys and apes. The first primates appeared about 55 mya, but looked nothing like us.

Plesiadapis
(PLEH-see-ah-dap-iss)

Primary primates

The first primates, like *Plesiadapis*, lived in the trees and looked rather like squirrels, with grasping hands and feet which helped them to climb. They are called **prosimians**.

Smilodectes (SMY-lo-deck-tees) was a prosimian that lived in the forests of North America around 50 mya.

From prosimian to simian

Some prosimians evolved in ways that eventually led to modern **simians**—monkeys, apes, and hominids (all types of human). Their pointed snout became shorter, and they stood more upright, with their head positioned above their back (as ours is) rather than in front of it, as in mammals like dogs and cows.

Dryopithecus (DRY-oh-PIH-theh-kuss) did not swing from trees, and could walk on all fours or upright.

Lemurs

Lemurs, tarsiers, and lorises are living prosimians. Lemurs live only in Madagascar, an island off the coast of Africa. Monkeys have never lived in Madagascar, so have never competed with the lemurs. Lemurs come out in the day there, while promisians elsewhere are nocturnal (they come out at night) when the monkeys are asleep.

Ring-tailed lemur

Floating monkeys

Monkeys split from prosimians about 34 mya. They spread to South America 30 mya, probably by floating on rafts of plants and soil ripped from the coast in violent storms. South America was an island, and closer to Africa than it is now. The monkeys evolved separately, having no chance to breed with those evolving in Africa and elsewhere.

Aegyptopithecus (EE-jip-toh-PIH-theh-kuss), an early old-world monkey.

From monkey to ape

Another split happened around 25 mya, when apes split from monkeys. Apes and monkeys are very different. Monkeys have a tail and walk along the tops of branches. Apes have no tail and swing from branches. Some animals that fall between monkeys and apes, with features of each, show evolution in action.

Gigantopithecus (gi-GAN-toh-PIH-theh-kuss), the largest ape ever.

Proconsul (PRO-CON-sul) was an early ape with monkey-like features that lived in Africa 23–25 mya.

ISLAND ISOLATION

The lemurs flourished in Madagascar because there were no monkeys to compete with them. Other lands that were cut off as islands have similar stories of species thriving and evolving independently.

The island of South America

The land that carry the continents moves less than a hand's width a year, so it takes a long time to tear a continent apart. At some point, the gap between lands becomes too large for animals to jump over, then too large to swim across, and finally too large for most animals to fly over.

Until about 30 mya, South America was linked to Antarctica, and Antarctica was linked to Australia. After it separated from Antarctica, South America was a huge island covered with tropical forest. Some of the strangest animals of the time evolved in isolation on that island.

Map of the world as it was 50 mya.

Macrauchenia, (MAH-krow-KEE-nee-ah) about 3 m (10 ft) long, had nostrils between its eyes, possibly a trunk, three-toed hoofs on strong, sturdy legs—and a body like a llama. It was a strange mix of features found in animals from rhinos to llamas and tapirs.

Phoberomys (foh-BEH-roh-miss) was like a giant guinea pig—really giant at 4.5 m (15 ft) long and 1.5 m (5 ft) tall. It ate grass, and probably spent some time in rivers, where giant crocodiles were a danger.

Megatherium (MEH-gah-THEE-ree-um) was a giant ground sloth, 6 m (20 ft) tall.

Argentavis (AR-gen-TAH-viss), the largest flying bird ever to live, had a 6-m (20-ft) wingspan and soared above the Andes.

Glyptodon (GLIP-toh-don) was a giant armadillo-like animal that shambled over the grasslands. The size of a small car, it had a shell of bony plates over its whole body, which protected it from predators.

Thylacosmilus (THY-lah-koh-SMY-luss) was a marsupial with dagger-like teeth. It looked rather like *Smilodon* (SMY-loh-don) of North America, but this was convergent evolution at work—as a marsupial, *Thylacosmilus* evolved from a very different line of mammals.

LIVING NOW
Capybara

The capybara now living in South America is a much smaller version of *Phoberomys.* It grows to only a quarter the length of its early ancestor. Capybara live near rivers in the dense forest, but can also live on grassland.

Capybara

Paraphysornis (PAH-rah-fih-SOR-niss) was a 2-m (7 ft) tall "terror bird"—huge, savage birds that lived in South America from around 23 mya. Their massive legs were adapted to running fast and kicking hard. Modern ostriches in Africa have evolved in the same way.

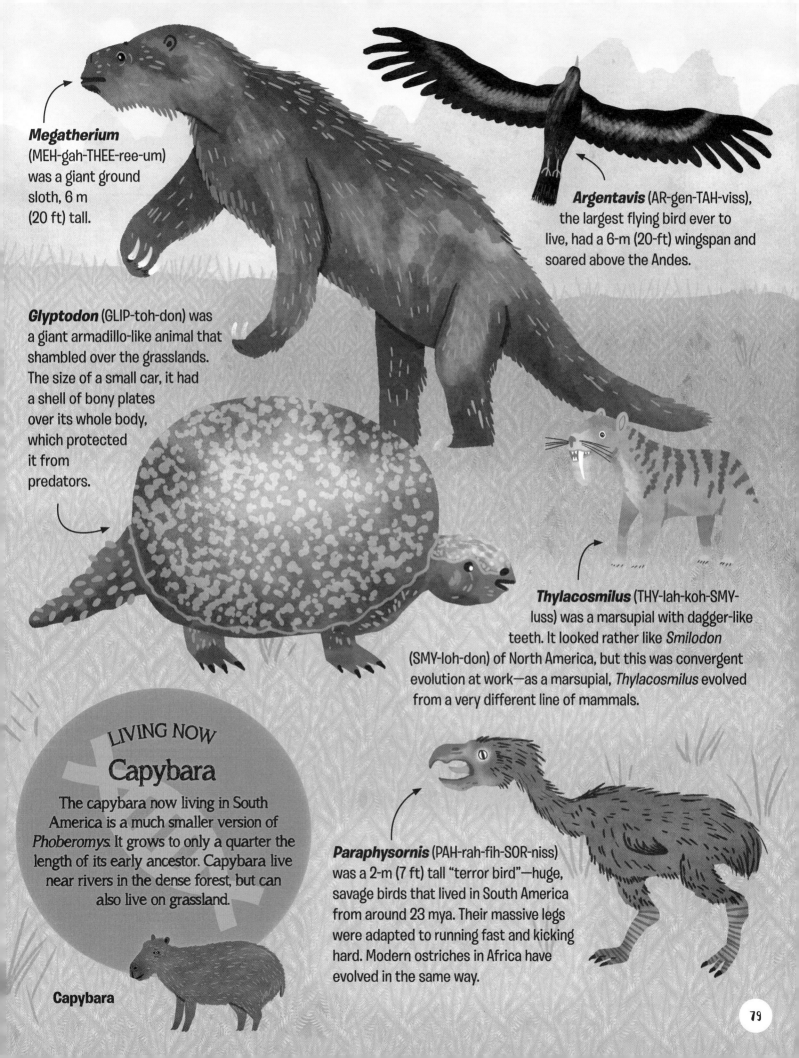

CHANGING PLACES

The monkeys that arrived in South America from Africa evolved independently of those in North America, as the two continents were not yet joined. South America was an island. It began to join to North America around 14 mya.

Joining lands

At the same time as South America drifted northward, volcanic activity pushed up land from the seabed, making a shifting pattern of islands and land between the two continents. At first, some species island-hopped, or were carried by wind or sea. When a full land bridge formed, 2–3 mya, animals from both continents swapped places in an event called the **Great American Biotic Interchange**.

All change!
More animals successfully moved from North America to South America than the other way round. South America was a continent of tropical rainforest. The animals that left for North America were used to living in tropical forest, but going north they soon found land that was too cold or dry for them.

Koala

KOALAS FROM SOUTH AMERICA?

Marsupials—mammals that keep their undeveloped babies in a pocket—first evolved in South America and spread from there to Australia, crossing ice-free Antarctica. They evolved separately in Australia and were successful. All the most familiar marsupials now live in Australia, including koalas, wombats, and kangaroos. Only a few still live in South America.

Kangaroo

The Virginia opossum lives in North America.

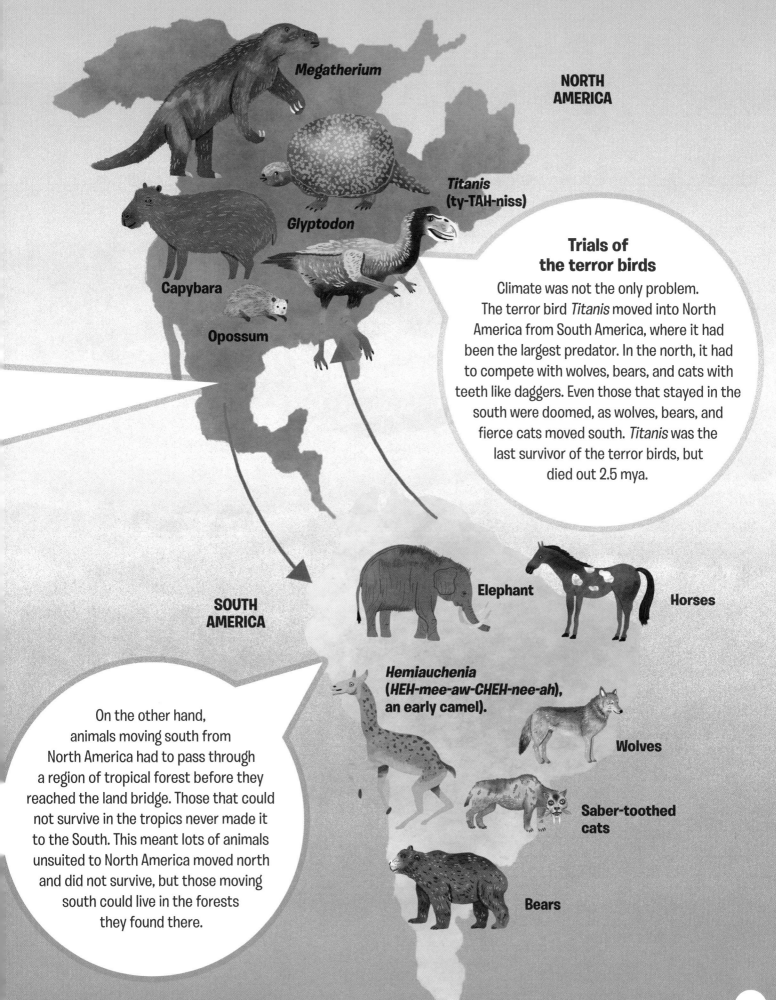

NORTH AMERICA

Megatherium

Glyptodon

Titanis
(ty-TAH-niss)

Capybara

Opossum

Trials of the terror birds

Climate was not the only problem. The terror bird *Titanis* moved into North America from South America, where it had been the largest predator. In the north, it had to compete with wolves, bears, and cats with teeth like daggers. Even those that stayed in the south were doomed, as wolves, bears, and fierce cats moved south. *Titanis* was the last survivor of the terror birds, but died out 2.5 mya.

SOUTH AMERICA

Elephant

Horses

Hemiauchenia
**(HEH-mee-aw-CHEH-nee-ah),
an early camel).**

Wolves

Saber-toothed cats

Bears

On the other hand, animals moving south from North America had to pass through a region of tropical forest before they reached the land bridge. Those that could not survive in the tropics never made it to the South. This meant lots of animals unsuited to North America moved north and did not survive, but those moving south could live in the forests they found there.

THE AGE OF HUMANKIND

Because we are human, we have a particular interest in the evolutionary path that led to humans. However, from the point of view of natural selection, there's nothing "special" about humans.

If we were sponges or ants, tapeworms, or lions (and capable of writing and reading books), we would prioritize a different evolutionary path. Humans are special in one way, though. We have had a huge impact on the environment and the evolution of other species—though still not as big an impact as cyanobacteria or mosquito fern.

People have recorded their interactions with nature for thousands of years, since they began painting the animals around them on the walls of the caves they lived in. They left not just fossils, as other organisms did, but pictures, tools, bits of their homes and their meals, and later written accounts and finally photographs, from all of which we can learn about the development and impact of our own species.

FROM MONKEYS TO HUMANKIND

The first of our long-ago ancestors to begin to look and act like we do were the early apes. They evolved around 25 mya, splitting off from the African monkeys.

Down from the trees

Apes spend much more time on the ground than monkeys. They are larger, and so safer from ground-based predators than monkeys. When they do take to the trees, they have shoulders adapted to swinging from branches, with their bodies hanging down, rather than walking along the tops of the branches as monkeys do. On the ground, they walk on their hind feet and the knuckles of their "hands." Most apes eat leaves, bark, and fruit with just the occasional insect or other small animal. They sleep in nests of leaves on the ground or in trees. Physically, they differ from monkeys in having a flatter face and no tails. They are larger and have larger brains.

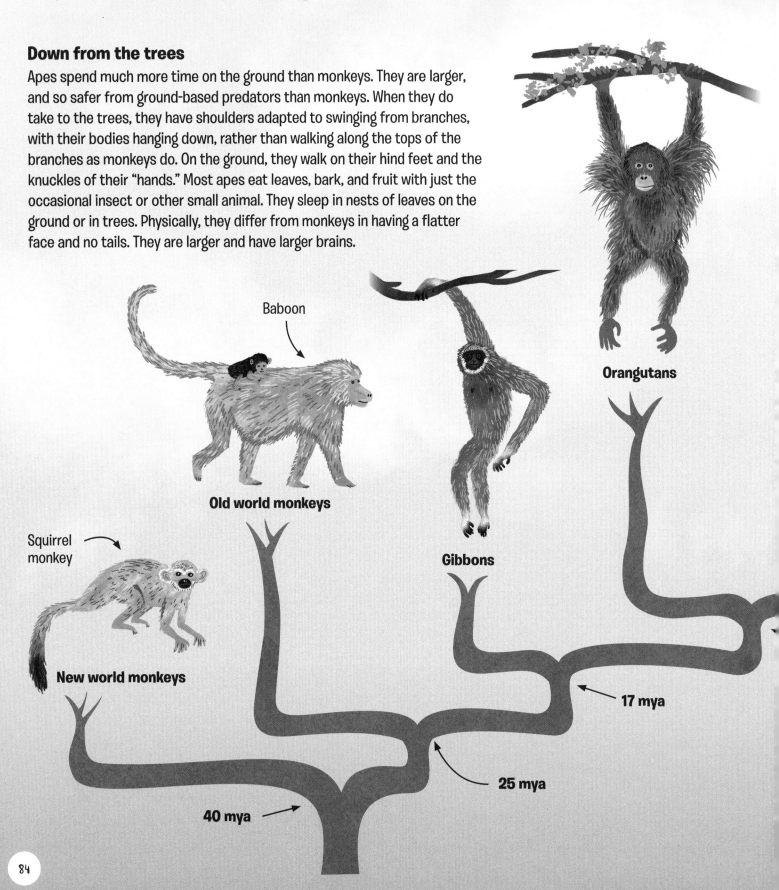

Orangutans

Baboon

Old world monkeys

Gibbons

Squirrel monkey

New world monkeys

17 mya

25 mya

40 mya

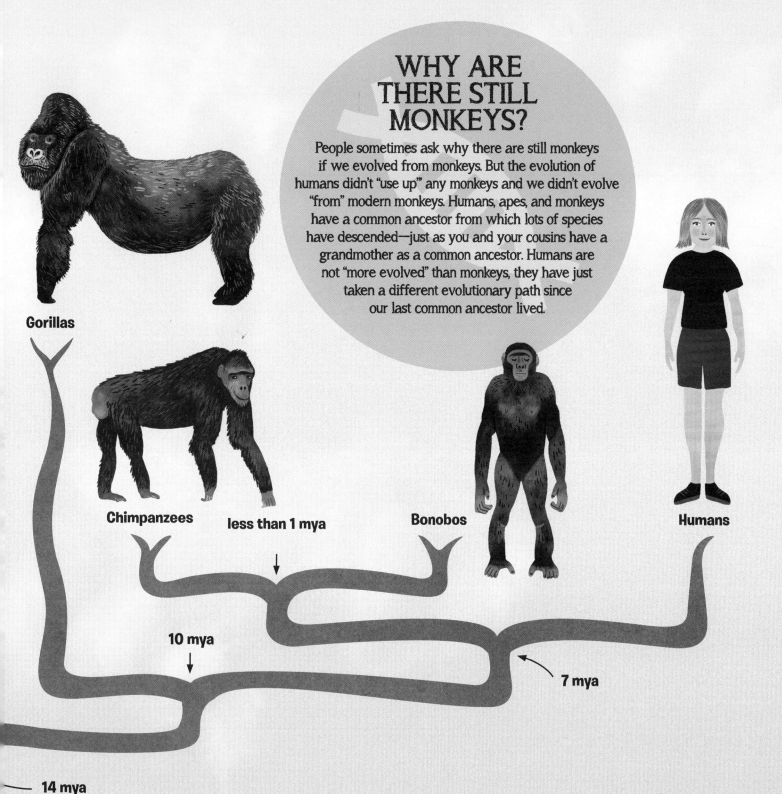

WHY ARE THERE STILL MONKEYS?

People sometimes ask why there are still monkeys if we evolved from monkeys. But the evolution of humans didn't "use up" any monkeys and we didn't evolve "from" modern monkeys. Humans, apes, and monkeys have a common ancestor from which lots of species have descended—just as you and your cousins have a grandmother as a common ancestor. Humans are not "more evolved" than monkeys, they have just taken a different evolutionary path since our last common ancestor lived.

Gorillas

Chimpanzees

less than 1 mya

Bonobos

Humans

10 mya

7 mya

14 mya

Apes, big and small

Apes now fall into two groups, called greater and lesser (smaller) apes. The lesser apes live only in Asia, but the great apes live in Africa and Asia—and of course humans live everywhere. Gibbons are all lesser apes; orangutans, gorillas, bonobos, chimpanzees, and humans are great apes. Humans are more closely related to chimpanzees than to the other great apes. Our evolutionary path split from that of the chimpanzees about 7 mya, when we last had a common ancestor.

THE FIRST HUMANS

Although we are the only species of human now living, we are not the only species there has ever been. Some came before us, and some lived alongside us for a while.

Becoming human

Important differences between humans and other apes made us successful. We walk upright on two legs, which frees our hands to use tools. We have larger brains than other apes. And our hands are a different shape, better for using tools than climbing in trees. Early humans probably spent most time on the ground.

All types of human are hominins. The earliest almost-human, *Sahelanthropus*, first walked on two feet about 7 mya. The first real hominin was *Australopithecus afarensis*, 3 million years later. It shared characteristics with non-human apes, such as a flattened face and a small brain, but walked upright and had teeth like ours.

Ardipithecus
(AR-dih-PIH-theh-kuss)

Homo ergaster
(HOH-moh er-GASS-tuhr)

Australopithecus afarensis
(OSS-trah-loh-PIH-theh-kuss
AH-fah-REN-siss)

Sahelanthropus
(SAH-hey-LAN-throw-puss)

Homo heidelbergensis
(HOH-moh HY-duhl-burg-EN-siss)

Homo neanderthalensis
(HOH-moh NEE-an-dur-tal-EN-siss)

Homo sapiens
(HOH-moh SAY-pee-enz)

Homo habilis
(HOH-moh HAH-bih-liss)

Homo floresiensis
(HOH-moh FLOH-reh-see-EN-siss)

EVOLUTIONARY PUZZLES
Homo Floresiensis

One species of early human, Homo floresiensis, was unusually small. Living on the island of Flores in Indonesia 190,000–50,000 years ago, these humans were just over 1 m (3 ft) tall. They used stone tools and possibly fire, and successfully hunted small elephants. No one knows why they were so tiny.

Homo arrives

Modern humans are *Homo sapiens*. Earlier *Homo* species were able to do some of the things we do. *Homo habilis* appeared 2.4 mya in Africa, looking recognizably human. They used stone tools, but the oldest stone tools are even older, so they weren't the first. *Homo erectus* was the first to cook meat, and lived from 1.95 mya until just 143,000 years ago.

Homo heidelbergensis, 700,000 years ago, were the first hominins to move into northern Europe, to build shelters, and to use wooden spears to hunt large animals. They were the ancestors of *Homo neanderthalensis*, who lived alongside *Homo sapiens* for 160,000 years, and died out just 40,000 years ago. Neanderthals were the first to wear clothes, bury their dead, and possibly use language. Their bodies adapted to the colder climate of Europe, with shorter fingers and toes to conserve body heat and a large nose to warm the air going into their lungs.

MIX AND MATCH— MAKING NEW SPECIES

Humans survive in only one species, but many it is possible to have several species living at the same time. Speciation—splitting into different species—can happen in several ways.

Spreading out

Imagine a large expanse of sea or land with few organisms. This happens after an extinction event. Organisms spread out, finding different places to live. They face different challenges of climate, landscape, and predators. Although they started out the same, they adapt differently to suit their new environments. They become separate species.

Drifting apart

Sometimes, a group of organisms will be split in two. Perhaps a river will change its course to run through their territory. Or one group might move elsewhere. The groups will develop differently. Each group has its own gene pool—the collection of all the genes available in the group. Some features will become more common in one group than in the other over time, just by chance. This is called genetic drift. Eventually, the groups might be so different that they are separate species.

Round in a ring

Groups of organisms close together might differ only slightly from their closest relatives. But small changes build up. The changes can be enough that those at each end can't interbreed. If the row of species forms a loop, and the individuals at each end can't breed, it's called a ring species.

oregonensis

picta

xanthoptica

eschscholtzii

Speciation of salamanders (*Ensatina eschscholtzii*) in California

platensis

crocreator

klauberi

88

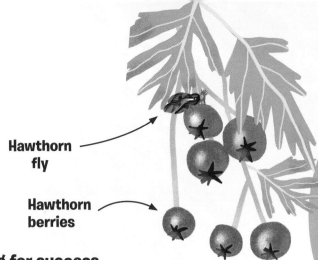

Hawthorn fly

Hawthorn berries

ADAPTATION IN ACTION

Hawthorn flies

Hawthorn flies originally fed on the berries of the hawthorn tree in North America. After settlers from Europe introduced apple trees, the species began splitting into two, with one set of flies feeding on apples instead. The arrival of a new food led to speciation.

Hawthorn fly

Selecting for success

Changes happen over time in any group. Some organisms have features better suited to their environment than others. They are more likely to survive to adulthood and breed, so their good features are passed on; this is the process of natural selection.

Many animals choose mates to breed with. They choose mates that are successful and seem strong and healthy. It sounds harsh, but an animal with an injury or a mutation that causes a disadvantage, or that is smaller than the others, is less likely to find a mate, so genes for strength and success are passed on. This is sexual selection.

Look at me!

Sexual selection can work in strange ways. Female peacocks choose males with large, bright tails. But a huge, bright tail probably makes a bird more easily seen and caught by predators. It seems that the message of the tail is, "I am SUCH a strong/fast/excellent bird that EVEN WITH this tail I am successful." The huge tail, a disadvantage in daily life, becomes an advantage when seeking a mate, so over time tails grow bigger.

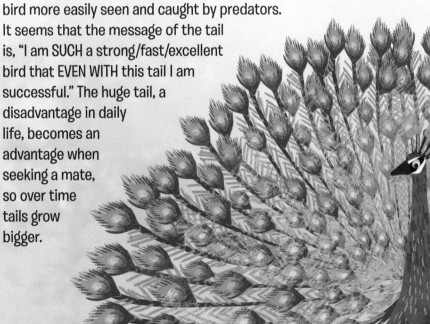

Peacock

OUT OF AFRICA

Modern humans, *Homo sapiens*, evolved in Africa around 200,000 years ago, but began to leave around 120,000 years ago. Humans spread first into the Middle East, then into South East Asia, North Asia, and Europe. They arrived in Australasia 50,000 years ago and in the Americas only 15,000 years ago.

Suited to sunshine

As modern humans evolved in Africa, their bodies were well adapted to hot, bright sun. They had dark skin, because that's best suited to being in the sun. The dark pigment, melanin, protects the skin from sun damage.

Our bodies make vitamin D only in sunlight. In Africa, the sun is bright enough that we make vitamin D easily. Further from the equator, where there is less sunlight, it's more difficult for dark-skinned bodies to make enough vitamin D. When humans moved north, people with slightly paler skin were healthier, so over time, the population grew paler. Over time, people near the equator kept dark skin, avoiding sunburn, and people elsewhere developed paler skin that could make vitamin D with less sunlight.

Europe
45,000
years ago

Levant and Arabian peninsula
120,000 to
90,000
years ago

Homo sapiens in Africa
150,000 to
200,000
years ago

NEANDERTHALS AMONG US

Usually, organisms can breed only with individuals of their own species, but sometimes closely related species can breed together. Lions and tigers can breed in zoos, and horses and donkeys can breed together (though their babies are not able to reproduce). *Homo sapiens* and Neanderthals could and did interbreed. We know this as gene analysis shows that modern people have a small percentage of Neanderthal genes still. The genes have been carried forward in *Homo sapiens* even though Neanderthals themselves died out 40,000 years ago.

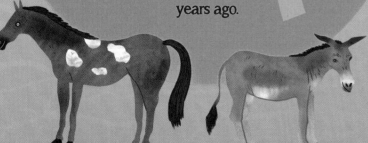

Horse **Donkey** **Mule**

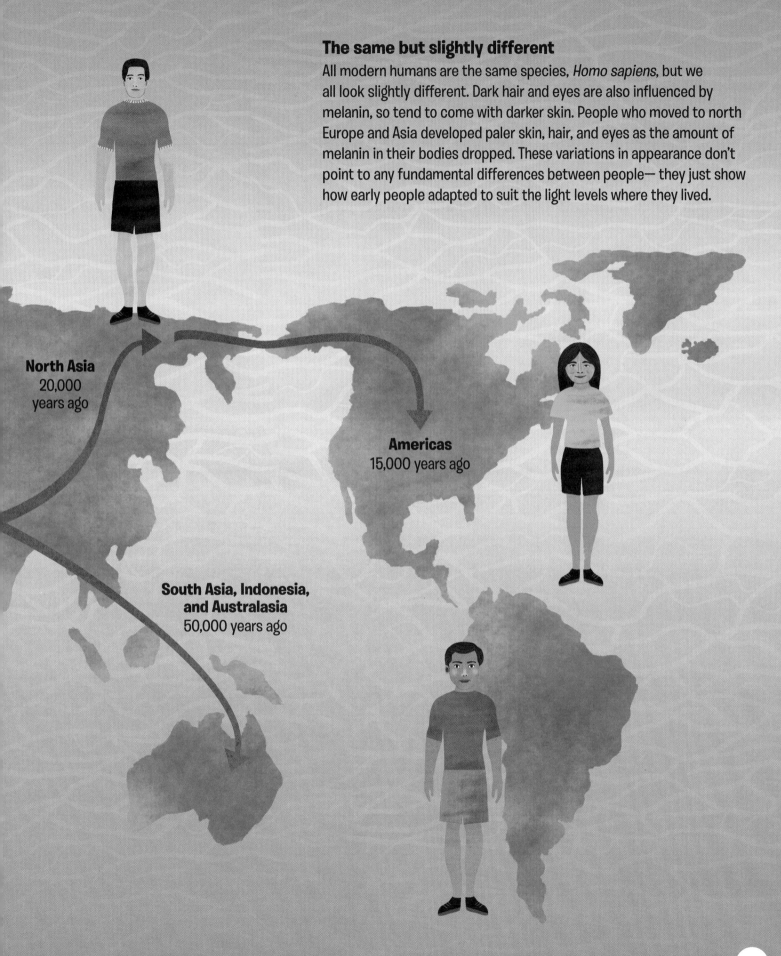

The same but slightly different

All modern humans are the same species, *Homo sapiens*, but we all look slightly different. Dark hair and eyes are also influenced by melanin, so tend to come with darker skin. People who moved to north Europe and Asia developed paler skin, hair, and eyes as the amount of melanin in their bodies dropped. These variations in appearance don't point to any fundamental differences between people— they just show how early people adapted to suit the light levels where they lived.

North Asia
20,000
years ago

Americas
15,000 years ago

South Asia, Indonesia, and Australasia
50,000 years ago

GOOD PARENT, BAD PARENT?

Evolution affects the ways that animals behave as well as their bodies—such as whether they take only one mate or several, or whether they look after their young. It's often difficult to work out how extinct animals behaved because most actions don't leave evidence.

How many children should you have?

To be successful, an organism needs to pass on its genes to the next generation by reproducing. And the young must survive, to pass on the genes again. There's more than one way to do this. An organism can have lots and lots of young, so there's a good chance at least some will survive—most plants do this, producing lots of seeds. Or an organism can have a small number of young and invest time and energy in caring for them, making it more likely they will survive.

Tadpoles

Sink or swim!

Plants can't look after their offspring, but many animals can. Some still don't! A frog produces frogspawn and leaves it. There are many eggs, and those that aren't eaten grow into tadpoles. The tadpoles are a welcome snack for many other animals, including fish, water birds, and reptiles. Not many survive to be frogs, and tiny frogs are vulnerable, too.

Nurturing in nature

Some other animals take more care of their eggs. Tilapia fish carry their eggs in their mouths to protect them from being eaten. Birds are the most careful of egg-layers, building a nest and sitting on the eggs to keep them warm until they hatch.

Tilapia fish

A dinosaur nest site

At least some dinosaurs cared for their young. Fossilized nests of *Maiasaura* from 77 mya include young animals, showing that they stayed in the nest after hatching.

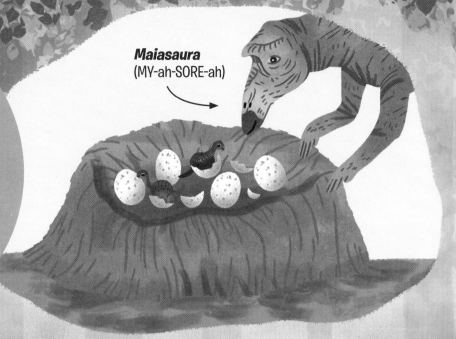

Maiasaura
(MY-ah-SORE-ah)

Some chicks are dependent on a parent, like these **hoopoes**.

On your own two (or four) feet

Some young need more care than others after birth or hatching. Birds are divided into those that hatch with feathers and can immediately walk and peck, and those that hatch without feathers and must be fed by their parents.

Some chicks feed independently as soon as they hatch, like **chickens**.

Caring parents

All mammals make milk for their young, but some do much more besides. Young deer, cows, and horses are wobbly to start with but can soon walk and run. Others have helpless infants. Cats and dogs are born blind. Some young stay with their parents for years (as humans do). These parents invest a lot in a few young.

Elephants make milk for their young for up to 4 years.

SHARING THE LAND

Humans have always shared the world with other organisms, from the microbes that live in and around us to giant trees and insects, mammals, birds, and fish. Humans are unique in the ways they have interacted with other species.

Creatures and climate

In Northern Europe and Asia, and North America, people encountered large animals (called megafauna) such as cave bears, *Smilodon*, woolly rhinos, giant beavers, giant deer, and the elephant-like mastodons and woolly mammoths. In Australia there were giant kangaroos, a marsupial lion, and *Diprotodon* (dy-PROH-toh-don), a wombat the size of a hippo. In the Americas, humans lived alongside giant birds, *Glyptodon*, and giant ground sloth.

Argentavis
(AR-gen-TAH-viss)

Gone forever

Many of these species died out around 12,000 years ago, having survived for millions of years. Human hunters have driven lots of megafauna to extinction, though climate change might have played a part, too. The climate has swung between glacial periods when ice has spread far from the poles, and warmer periods when even the poles were ice-free.

Megatherium
(MEH-gah-THEE-ree-um)

Smilodon
(SMY-loh-don)

Humans interacting

Early species of human hunted many of the large animals, but that was not all they did. They lit fires and cleared ground, changing the environment in a way no previous species did. Humans used the skins of their prey to make simple clothes, which meant they could live in colder places than their bodies were adapted to. They made tools from the bones of their prey and even carved or painted images of them, leaving evidence of the animals they lived among.

Glyptodon
(GLIP-toh-don)

Too hot to trot

The megafauna that were suited to a cold climate, such as the woolly rhino and mammoth, might have died when the weather got too warm. Animals that need a warmer climate can move toward the equator during glacial periods, but animals that need a cold climate have nowhere to go in warm periods. If some plant-eaters die out, predators and carrion-eaters that depend on them often die, too.

HUNGRY HUMANS

Early humans would have eaten meat and fish they caught, fruit, seeds, nuts, and other plant matter. They had none of the fruit and vegetables we know now, as all of these have been changed by being selectively bred by farmers over thousands of years.

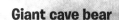
Giant cave bear

95

EVOLUTIONARY ARMS RACE

Some species co-evolve, helping each other along, but others are caught in a battle to live, each trying to outdo the other in a war for survival.

Predator v. prey

Many predators and prey just keep on doing the same things, but harder, faster or better. The whelk has a hard shell, but crabs have evolved strong claws to crack the shell. So the whelk's shell is getting even stronger—but so are the crab's claws.

Hoary bat

Harnessed tiger moth

Lepidodendron

Sneaky tricks

Some predators or prey animals evolve camouflage—feathers, fur, scales, or skin in hues and patterns that hide them in their environment, such as flatfish that blend into the sandy seabed or polar bears that are hard to see in the snow. Others develop new tricks. Bats hunt in the dark using echolocation; they make high-pitched sounds and can tell from the echo where their prey is. But the North American harnessed tiger moth has evolved to block the echolocation of the bats that eat it.

Space invaders

Trees compete with each other for space and light. Even 300 mya, trees like *Lepidodendron* and *Synchsidendron* competed with each other, each growing taller and taller before spreading their branches.

Cuckoo chick

A cuckoo in the nest

Cuckoos lay their eggs in the nests of other birds, cheating the hosts into hatching and raising the cuckoo chicks. Some birds have evolved defences against cuckoos—throwing out eggs that look too large, for example. In Zambia, it's got more serious. Cuckoo finches have evolved to look like common, harmless birds so that the host birds don't notice them. But now the hosts have started to scare away anything that looks like the harmless bird.

Organism v. disease

Microbes evolve quickly, including those that cause diseases. They can keep up with changes in organisms they affect, and even with the treatments and medicines we produce. When a mutation changes a microbe so a medicine no longer works against it, that microbe survives and reproduces, making more and more resistant microbes.

Bannaquit bird

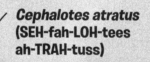
Cephalotes atratus
(SEH-fah-LOH-tees
ah-TRAH-tuss)

Parasite v. host

Parasites live on or in another organism (their host). They can be as tiny as a microbe or as large as a 30-m-(98-ft-) long worm. The parasitic plant mistletoe grows on trees, taking water and food from its host.

Some parasites change the appearance of their host or the way it behaves. A parasitic worm that lives in a kind of tropical ant changes the ant's body to look like a red berry. The "berry" body attracts birds. When a bird eats the ant, the parasite passes out in the bird's droppings, and is picked up by other ants feeding on seeds in the droppings.

Parasite inside ant's body

AGENTS OF CHANGE

Although many organisms change the environment, few do it as much as humans. For thousands of years, we have changed the landscape by clearing land, blocking rivers, and changing species through farming and domesticating (taming) animals and plants.

Home from homes

Early humans developed stone axes to cut down trees and used fire to clear away scrub and undergrowth, destroying the habitats of other organisms. From around 12,000 years ago, people built homes and farmed crops and animals on land they cleared. As they farmed, they irrigated the land, building channels to control the flow of water. They removed unwanted plants and scared away or killed animals that might eat their crops or herds. This reduced the diversity of wildlife where people lived.

Changing lands

Clearing land of trees changes the weather and how water runs over the land. This affects other organisms. Farming changes the conditions of the soil and affects the organisms that live in it. When just one kind of plant is grown instead of many, the animals that feed on plants must evolve to change what they eat, move somewhere else, or die out.

Changing animals

People keep animals to provide meat, milk, and wool or hide, or to help us hunt or keep away pests and predators. We don't keep animals that are no use to us. No one farms giraffes or woodpeckers, because these don't provide anything people use. People began to keep sheep, goats, cows, and pigs—and then we began to keep dogs (starting as wolves) to protect our animals from predators. The domesticated animals began to change, evolving to suit their new lifestyles.

Feeling sick?

When humans began to live and work closely with animals, some of the microbes that caused diseases in those animals evolved to infect humans. Human diseases that began in animals include smallpox (from rodents), measles (from cattle), bubonic plague (from rodents), and flu (from birds and pigs).

Slowed by sickness

Ten thousand years ago, there were 4 million people in the world. Five thousand years later, the population was still only 5 million. (It's now about 8 billion, or 8,000 million.) During this period, people around the world began to farm and build communities. At the same time, disease and wars kept the population in check. When people lived in larger groups in fixed locations, diseases could spread more easily than when they lived in small groups, moving around. Disease-causing microbes co-evolved with human societies.

EVOLUTION ALL AROUND US

Over the last 200 years, people have come to understand evolution and how it works. Recently, we have even gained something no other species has ever had—the ability to influence evolution deliberately in some ways. For many species, our impact has been negative.

Humans have destroyed habitats and environments, driven species to extinction, and reduced the diversity of the natural world. We have influenced the evolution of some species, shaping crops and domesticating animals to suit our own purposes. As our understanding of genetics has improved, we have gained the ability to create entirely new organisms, mixing genes from different species and fixing genetic problems rather than waiting for evolution to do its work. Will we put this to good use?

A BREED APART

The animals and plants that we keep and grow haven't evolved naturally. We have changed them by choosing which ones to breed together to get offspring with the features we want.

Taming the wild beast

As soon as people began to keep animals and plants, we started to change them. We took wolves and tamed them, slowly turning them into dogs. And then we changed dogs into lots of different breeds (but they are all still the same species.) We kept animals for milk, eggs, and meat, choosing the woolliest or largest or best egg-layers and breeding from those. This is called "selective breeding."

Mouflon (wild sheep)

Top crops

With plants, a farmer keeps some of the seed from the best crops and, instead of eating it, plants it the following year. That way, the whole crop can have the features of the best plants.

With animals, farmers breed the best animals together. A farmer could breed from the two woolliest sheep, and then from the woolliest offspring, and so on. This strengthens the genes for woolliness in the flock and over time, the sheep become more woolly. The sheep are no longer selected by their ability to survive, but by humans for their usefulness. We have kidnapped evolution and made it work for us.

Modern domestic sheep

Teosinte (wild maize)

Modern maize

Grain gains

None of the modern fruit, vegetables, and grains we enjoy are in their wild state. They have all been selectively bred over hundreds or thousands of years to be bigger, juicer, tastier, and sweeter. The original maize plant looked more like wheat, and the original wheat more like grass. Carrots were once spindly, tough roots, and fruits much smaller.

Easter eggs?

Modern chickens lay eggs all year round, but that wasn't how the birds originally lived. They have been selectively bred until they produce eggs all year. This would not be an advantage for birds in the wild, which hatch their eggs in the spring so that their chicks have a summer of warmth and plenty of food before winter comes.

Good for us

What is good for us is not always good for the organism we're farming. A modern sheep produces too much wool to be healthy in the wild. Some dog breeds have flattened faces that make breathing difficult, or legs so short they can't run. The fruit and vegetables we grow are larger than they need to be to reproduce successfully.

Modern carrot

Wild strawberries

Modern strawberries

Wild carrot

103

ALL TOGETHER NOW!

Evolution doesn't only work at the level of individuals. Some organisms work together in groups. This is how multi-celled organisms first began—as groups of cells living and working together, then specializing to carry out particular tasks within the community of cells.

Social animals

As humans settled in larger groups, they split up tasks between them. Instead of each person hunting or picking their food, cooking it, making their own weapons and clothes, treating their sick, and teaching their children, people specialized. Some people farmed, others made tools or clothes, and so on.

Some other animals also split up roles and have gone further than humans, evolving physically different bodies suited to their tasks. As a person, you can choose whether to be a doctor, farmer, or actor—a bee is born to be a worker, drone, or queen and has no choice. Its body is suited to only one function.

One for all ...

Animals that work as large communities are called eusocial animals. Some kinds of ants, termites, bees, and wasps are eusocial. They live in a communal nest and all work for the good of the whole community.

In a beehive, a single queen lays all the eggs and is the mother of the whole colony. Other females are sterile (they can't reproduce). They work as soldiers, defending the hive or workers, building the hive, collecting food, and looking after the young. All the males, called drones, fertilize the queen's eggs. They don't do any other work. Among termites, workers and soldiers live separately. The workers are blind, and work only in the nest, while the soldiers defend it against attackers.

Workers

Queen

Drone

Animal cities

Eusocial insects are among the most successful animals in the world. In South America, leaf-cutter ants eat more leaves than all plant-eating mammals. In the southwest of the United States, there's a greater mass of ants and termites than of all other animals put together. Colonies of tiny termites can contain millions of individuals, and some termite mounds are occupied for hundreds of years and become very big indeed!

Termite mound

ADAPTATION IN ACTION
Burrowing together

Most eusocial animals are insects, but the naked mole rat is a mammal. It lives in a complex system of underground burrows. A single queen and a small number of fertile males reproduce, and the others maintain the burrows, collect food, and look after the young. The working females are sterile, but not permanently. When the queen dies, one of the others "switches on" the hormones to become fertile and takes over the job.

HOME ON THE RANGE

The area where a species can live is called its range. Many organisms live in a very restricted range. Others have a very wide range—humans live all over the world.

Large tree finch

Small tree finch

Medium tree finch

These tree finches have a short, stubby beak to grasp insects.

Stay where you are!

Most land species have a restricted range, even if it's quite large. They are limited by climate, availability of food, geography (rivers, mountains, or oceans blocking the path to elsewhere), and the presence of predators or competitors in other regions. Some animals migrate to follow the climate or food they prefer. Flamingoes winter in Africa and fly north to Europe for the summer. They have two seasonal ranges and a path between them.

Island species are geographically trapped in an area and can't spread further. This can affect their evolution. The finches of the Galapagos Islands all evolved from a seed-eating ground finch from the mainland of South America. Finches flew or were blown to the islands, where they all evolved separately, each developing a beak suitable to a type of food available on the island. There are now 14 distinct species. Finches evolved similar beaks to eat insects or fruit or seeds, even though they were separated.

Keeping it local

Other species are limited by the food available or the climate. Cacti can live only in hot, dry places, so will not spread into cold or wet regions. Giant pandas need to eat bamboo, so they only live in the bamboo forests. As the climate changes, some organisms are forced to change their range or adapt. Cold-living species have to move further toward the poles. This puts pressure on the organisms already living in that area.

ADAPTATION IN ACTION
Warmer world

Global warming is changing the range of many organisms. It's not just that cold-loving species have to move further north to escape the heat; heat-loving species can also move into a wider range. Mosquitoes that carry diseases like malaria, which causes dangerous fevers, are moving up from Africa into parts of Europe where they could not previously survive.

Large ground finch

Small ground finch

Medium ground finch

Mangrove finch

Mangrove and Woodpecker finches have longer beaks to use sticks as tools to get at grubs hidden in trees.

Woodpecker finch

BEAKS AND BIRDS IN THE GALAPAGOS

Warbler finch

These finches have a narrow beak to probe for insects.

Vegetarian finch

This finch has a beak suited to eating buds, flowers, leaves, and seeds.

Cocos Island finch

Large cactus finch

Cactus finches pull out seeds and parts of cacti.

Common cactus finch

Sharp-beaked ground finch

Ground finches pick seeds from the ground.

FAST FORWARD

Organisms evolve when they are under pressure from changes in their environment. If they can't change quickly enough, they die out.

Pressure to change

The peppered moth from Britain and Ireland comes in two forms. Usually, it has pale, mottled wings, but a mutant form has black wings. Usually, there are more pale moths. Pale moths are hidden against the tree trunks they rest on during the day, and predators can't easily see them. The dark moths are easily seen against the light tree trunks and are more often eaten by predators.

Light peppered moth

Dark peppered moth

Light and dark

In the nineteenth century, soot from factories and coal fires blackened the trees. The pale moths became more visible, but darker moths were harder to see. Over a short period, dark moths became the most common in cities. The first darker moth was spotted in 1848. By 1900, most moths in the city of Manchester were dark. In the twentieth century, when trees were no longer blackened by soot, the pale moths became more common again.

UNPICKING EVOLUTION

By 1900, most moths in the city of Manchester were dark. Scientists looking closely at the genes of peppered moths have worked out what happened. A tiny bit of one gene moved in a rare mutation in a moth hatched around 1819. It was successful, and reproduced, making more dark moths. By looking at variations in the gene that changed, scientists worked out how many generations ago it mutated.

Going nowhere

When organisms are well-suited to the conditions where they live, they don't need to change and they can stay largely the same for a long time, or change slowly by genetic drift. The fern *Osmunda claytoniana*, found in East Asia and North America, has stayed unchanged for at least 180 million years.

Osmunda claytoniana (oz-MUN-dah clay-TONE-ee-ah-nah)

Evolution by creeps and jerks

Scientists disagree about whether evolution generally creeps along at a slow and steady pace or whether it stays still for a time and then suddenly jerks forward, making a big leap, and then stops again. Do organisms usually change gradually all the time, or suddenly change in big ways?

Finding fossils

When evolution is slow and steady, there's enough time for fossils to form of the different stages. When it's sudden, there are few intermediate fossils, leaving a gap in the fossil record. There are fossils from before and after the change, but not in between. For a long time, there were no fossils between fish and the land-going creatures evolved from fish. The fishapods are helping to bridge the gap.

109

FACING EXTINCTION

We are currently living through the sixth mass extinction event, and this time it has been caused by a single organism—humans. The rate at which organisms are going extinct is 100–1,000 times the usual rate between extinction events, and even 10–100 times as fast as in previous mass extinctions.

Introducing invaders

As well as the damage caused by hunting and clearing land, humans have moved organisms around the world, introducing them to new places where they have sometimes done terrible harm. A species brought in from outside an area and which then settles is called an invasive species. Often, invasive species could never have found their way to their new home without human help. Some invasive species drive local species to extinction.

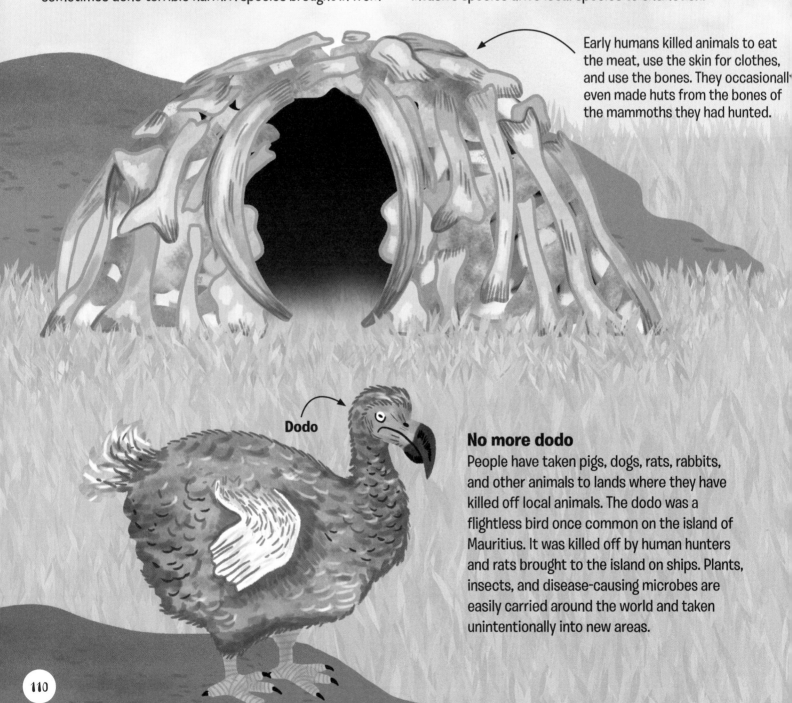

Early humans killed animals to eat the meat, use the skin for clothes, and use the bones. They occasionally even made huts from the bones of the mammoths they had hunted.

Dodo

No more dodo

People have taken pigs, dogs, rats, rabbits, and other animals to lands where they have killed off local animals. The dodo was a flightless bird once common on the island of Mauritius. It was killed off by human hunters and rats brought to the island on ships. Plants, insects, and disease-causing microbes are easily carried around the world and taken unintentionally into new areas.

Even being careful isn't always enough. Oysters from the Pacific Ocean have been farmed in the North Atlantic Ocean. As the sea there was too cold for the tropical oysters to live outside the farms, it seemed safe. But global warming has raised the sea temperature and now the Pacific oysters can survive in the open sea, threatening local species of oyster.

Western black rhino, driven to extinction in 2011 by hunting.

Wild dogs and farmers protecting sheep killed off the **Thylacine** (Tasmanian tiger).

Passenger pigeon, wiped out in 1914 by hunting.

Changing places

Humans have done harm in other ways, too, such as polluting the environment. Poisonous chemicals in the air, water, and soil kill other organisms. Other human activities can be just as bad. The Baiji freshwater dolphin, which lived in the Yangtze River in China, is probably now extinct—there were only 13 left in 1999. The dolphins have been accidentally killed by modern methods of catching fish. People have destroyed vast areas of rain forest and the coral reefs in the sea, both important habitats for hundreds of thousands of species.

EVOLVING TO SURVIVE

An accident in a nuclear power station in Chernobyl, Ukraine, in 1986 made the area around the reactor dangerously radioactive. It has been abandoned by humans, but plants and animals have moved back in. Despite the high levels of radiation, wildlife is flourishing. Some species are evolving the ability to live with radiation. A type of *Cladosporium* fungus now grows best on radioactive surfaces.

TAKING CONTROL

Humans can now intervene in evolution in new ways. We can target organisms to save or destroy, and can change the genetic make-up of others to suit our own purposes, going beyond anything that could turn up in nature.

Chopping and changing

Since the twentieth century, people have used special techniques called genetic engineering to change the DNA of organisms. Changing the DNA of a fertilized egg cell changes all the cells of the growing organism. We can use genetic engineering to produce organisms that could never occur naturally. One way of doing this is to take genes from different species that could never interbreed.

Designer genes

Scientists first changed organisms by moving particular genes between them. This can give an organism new characteristics. It's been used to create plants that don't get diseases, or that are better for us, or can live in harsher conditions. A genetically engineered form of rice, called "golden rice," is more nutritious than normal rice. Scientists have even made glow-in-the-dark mice that are useful for research.

A newer technique lets genetic engineers swap tiny bits of a gene. They might one day be able to use this to "fix" genes that have gone wrong to cure inherited diseases or stop cancers growing.

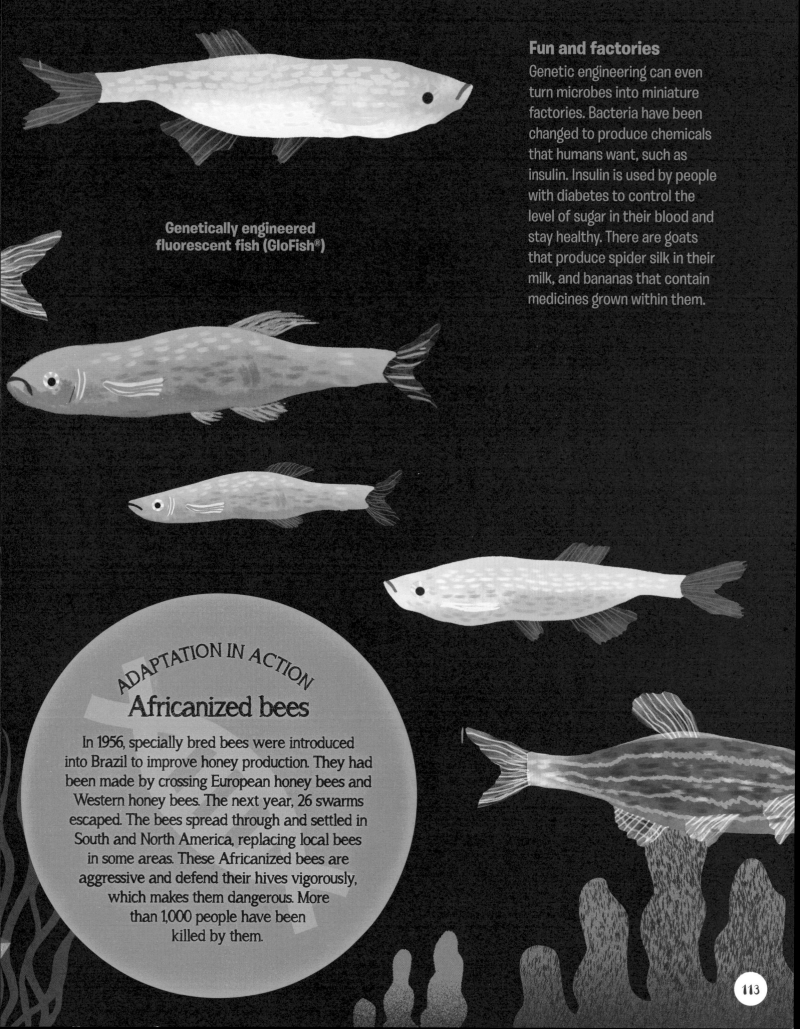

Genetically engineered fluorescent fish (GloFish®)

Fun and factories

Genetic engineering can even turn microbes into miniature factories. Bacteria have been changed to produce chemicals that humans want, such as insulin. Insulin is used by people with diabetes to control the level of sugar in their blood and stay healthy. There are goats that produce spider silk in their milk, and bananas that contain medicines grown within them.

ADAPTATION IN ACTION
Africanized bees

In 1956, specially bred bees were introduced into Brazil to improve honey production. They had been made by crossing European honey bees and Western honey bees. The next year, 26 swarms escaped. The bees spread through and settled in South and North America, replacing local bees in some areas. These Africanized bees are aggressive and defend their hives vigorously, which makes them dangerous. More than 1,000 people have been killed by them.

TURNING BACK THE CLOCK

With genetic engineering, we might be able to bring back species that have gone extinct. Whether we can and whether we should are two separate questions.

Stranger than fiction

The novel and movie *Jurassic Park* are based around the idea that we could make dinosaurs from their DNA. In the story, dinosaur blood is found in the guts of biting insects trapped in amber. Amber is a sticky resin from trees that hardens, sometimes with organisms stuck inside. Some bits of dinosaur (and insects) have been trapped in amber, but they couldn't be used to bring back dinosaurs. DNA decays too quickly to survive that long. There is no dinosaur DNA to use.

Some scientists in the USA are trying to make something like a dinosaur from chicken embryos. Chickens, like other birds, are related to dinosaurs. Their DNA still contains dinosaur features but these are turned off. By turning the genes back on, the scientists hope to grow chickens with tails, teeth, and other dino-features.

Pyrenean ibex

Woolly mammoth

Thylacine

THE BRIEF RETURN OF THE PYRENEAN IBEX

The last Pyrenean ibex died in 2000, killed by a falling tree. Before her death, scientists collected cells from her and produced a clone, which they grew in a female goat. The baby ibex was born, but died after 7 minutes because of a lung problem. It is still counted as the first de-extinction.

(Not) as dead as a dodo

It might one day be possible to bring back animals that have gone extinct much more recently than dinosaurs—as long as we have good-quality DNA from them. Perhaps we could bring back the woolly mammoth, thylacine, quagga, and even dodo. The woolly mammoth is closely related to the elephant but with important differences that make it suited to cold weather. It has a thick, shaggy coat, blood containing chemicals to tolerate cold, and small ears to reduce heat loss. By putting the genes for these features into elephant DNA, scientists hope to grow an elephant-mammoth with all the features of a mammoth.

Quagga

Back from the dead?

Even though we can't bring back extinct organisms yet, it's useful to think about whether it's a good thing to do. For scientists, it's a challenge and an opportunity to learn more about genetics and about the organisms themselves. But the organisms would be coming into a world where they would be alone and where conditions have changed. The conditions that possibly killed their kind might still be here. The food they eat might not be available and their habitat might have gone. Where would a twenty-first-century *Triceratops* live? Is it fair to bring these extinct species back? There are practical and ethical questions to consider.

WHERE NEXT?

Evolution hasn't ended and won't end, for as long as there is life on Earth. Even at the moment, humans and the other species on Earth are evolving all the time.

Coming back from catastrophe

The natural world has survived extinction events in the past and will no doubt do so in the future, whether they are caused by humans or by something else. We have seen life spring back after a complete change in the atmosphere, and in the climate; after catastrophic volcanic eruptions and at least one major asteroid strike. The climate change that we are seeing now, caused by human activity, might wipe out many species, possibly including ourselves, but life will bounce back. The world has been much hotter than it is now for most of the time that there has been life on Earth, though the speed of change is faster than ever before.

Perhaps we can fix the damage we have done, or slow it enough to stop a disaster happening. Then future people can watch as evolution takes its regular twists and turns.

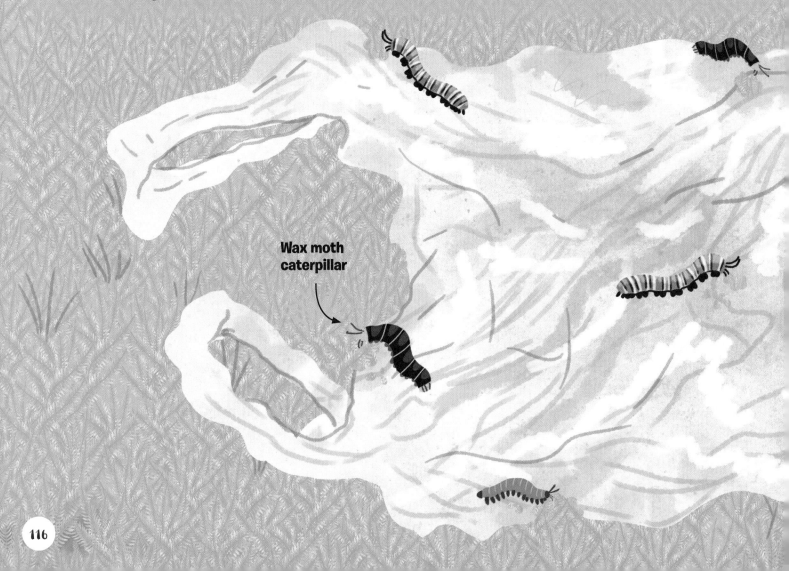

Wax moth caterpillar

ADAPTATION IN ACTION

Moths munching on plastic

The plastic waste humans produce is dangerous to many organisms, but perhaps some might evolve to use it. In 2017, researchers in Spain found caterpillars of the wax moth *Galleria mellonella* eating plastic bags. Scientists still need to confirm that they really do eat bags, and it would take a lot of caterpillars to destroy all our plastic waste—but evolution is all about adaptation.

The final frontier

Humans will change, too, adapting to fit our urban environments—or whatever replaces them—just as we adapted to living in the plains rather than the trees. If we ever colonize other worlds, the humans that are separated could even evolve into a new species—there might once again be more than one species of *Homo* alive at the same time.

EVOLUTION TIMELINE

From the earliest simple life forms, evolution has produced the millions of different types of plants and animals that live on Earth today. By changing and adapting, diversifying, and moving around, organisms have come to live in every corner of our planet.

Archaea, from at least 3.8 bya

Cyanobacteria, from 3.5 bya

GREAT OXYGENATION EVENT

Bangiomorpha, 1,200 mya

Prokaryotes

Bacteria are **prokaryotes**. They have a single, simple cell. There are still more bacteria on Earth than any other living thing.

Archaea were the first living things, with a simple, single cell. There are still archaea today, living in all kinds of environments around the world.

Eukaryotes

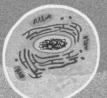

Bangiomorpha was the first organism made of lots of cells that grouped together. It was also the first to reproduce sexually—bringing two parents together to create more *Bangiomorpha*.

Cyanobacteria were the first things to photosynthesize—and the first to change the world. They filled the water and then the air with oxygen, changing conditions forever and causing the first mass extinction.

All the plants and animals around us today are **eukaryotes**. Perhaps the biggest step in evolution was the appearance of these more complicated cells. The first eukaryotes still had just one cell.

4 BYA → ← **3.5–2 BYA** → ← **2 BYA** → ← **1,200 MYA** →

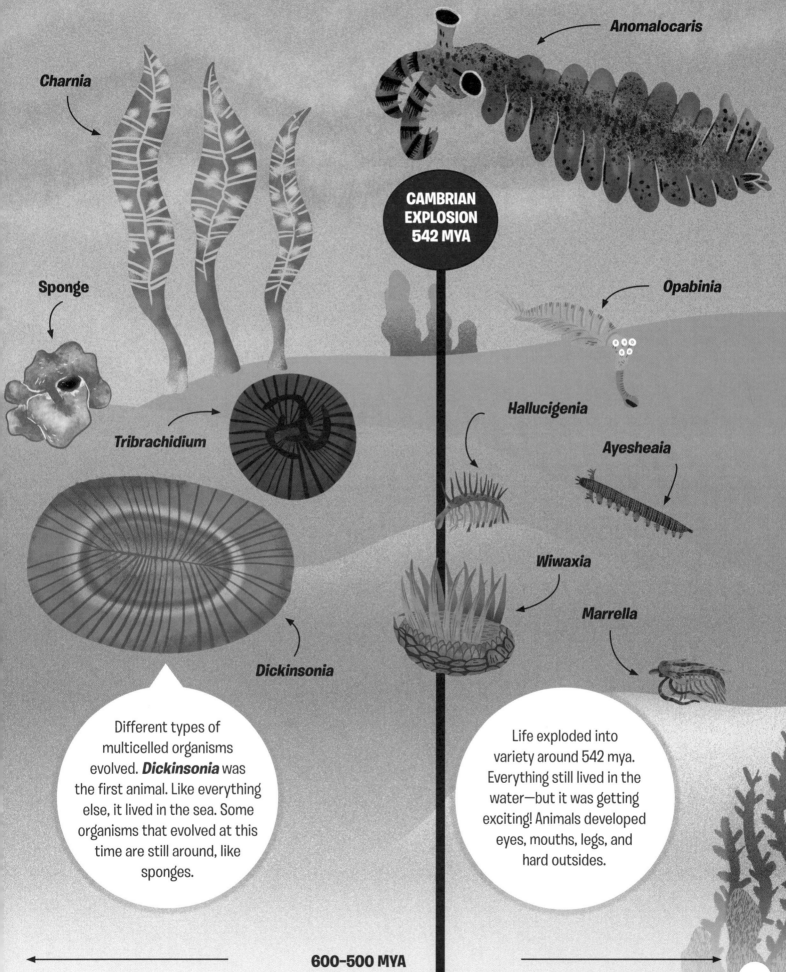

Charnia

Anomalocaris

Sponge

CAMBRIAN EXPLOSION 542 MYA

Opabinia

Tribrachidium

Hallucigenia

Ayesheaia

Wiwaxia

Marrella

Dickinsonia

Different types of multicelled organisms evolved. **Dickinsonia** was the first animal. Like everything else, it lived in the sea. Some organisms that evolved at this time are still around, like sponges.

Life exploded into variety around 542 mya. Everything still lived in the water—but it was getting exciting! Animals developed eyes, mouths, legs, and hard outsides.

600–500 MYA

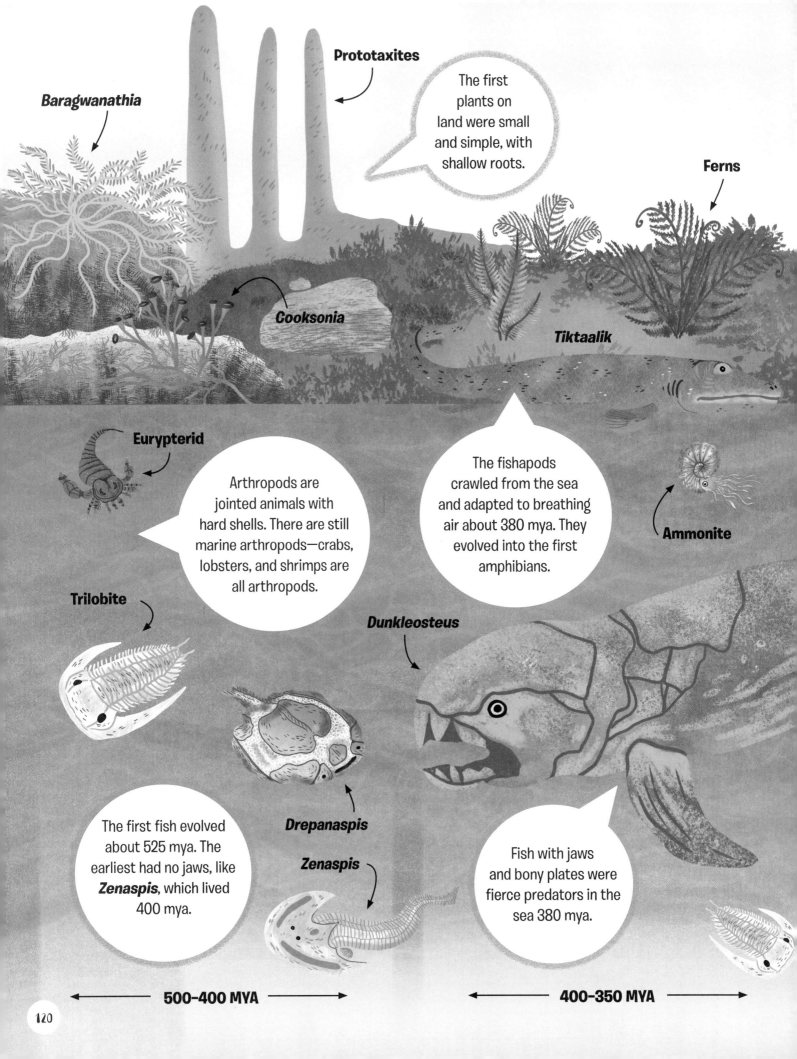

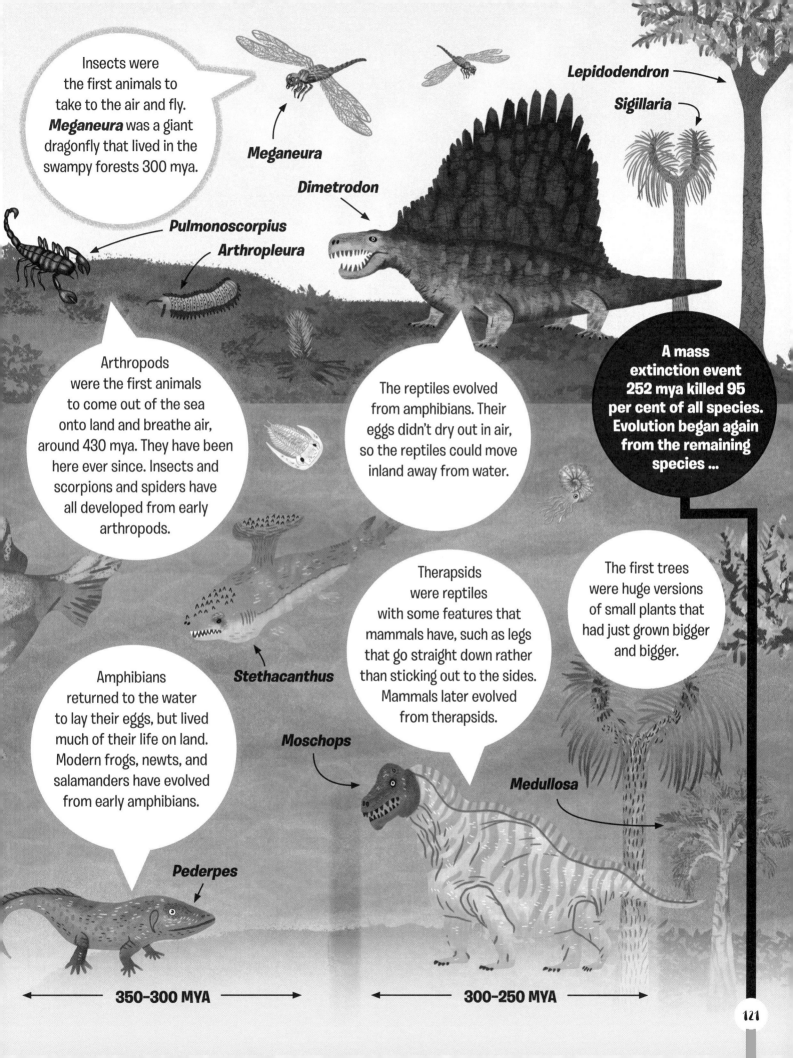

Insects were the first animals to take to the air and fly. **Meganeura** was a giant dragonfly that lived in the swampy forests 300 mya.

Meganeura

Dimetrodon

Lepidodendron

Sigillaria

Pulmonoscorpius

Arthropleura

Arthropods were the first animals to come out of the sea onto land and breathe air, around 430 mya. They have been here ever since. Insects and scorpions and spiders have all developed from early arthropods.

The reptiles evolved from amphibians. Their eggs didn't dry out in air, so the reptiles could move inland away from water.

A mass extinction event 252 mya killed 95 per cent of all species. Evolution began again from the remaining species ...

Stethacanthus

Therapsids were reptiles with some features that mammals have, such as legs that go straight down rather than sticking out to the sides. Mammals later evolved from therapsids.

The first trees were huge versions of small plants that had just grown bigger and bigger.

Amphibians returned to the water to lay their eggs, but lived much of their life on land. Modern frogs, newts, and salamanders have evolved from early amphibians.

Moschops

Medullosa

Pederpes

← **350–300 MYA** →

← **300–250 MYA** →

Starting from the survivors on land and in the sea, evolution repopulated the planet with new types of plants and animals.

Dimorphodon

Pterodactylus

The pterosaurs evolved from land-going reptiles about 220 million years ago.

Survivors of the extinction event spread and multiplied. **Lystrosaurus** was one of the most successful.

Lystrosaurus

As the land and climate changed, dinosaurs became more successful. Different types evolved—large, small, plant-eating, and meat-eating. They ruled the land for more than 150 million years.

Postosuchus

Herrerasaurus

Juramaia

Archosaurs were reptiles that later evolved into dinosaurs, crocodiles and birds.

The first dinosaurs evolved from earlier reptiles to fill some of the gaps opened up by the extinction event. They were small and agile.

The first mammals were small. They lived in trees and in burrows underground where they were safe from larger animals.

Ichthyosaurus

Liopleurodon

Cymbospondylus

Some reptiles returned to the water. They still had to breathe air, but their bodies adapted to swimming.

Some marine reptiles evolved to look like fish, with limbs that looked like fins.

← **250–200 MYA** →

← **200–150 MYA** →

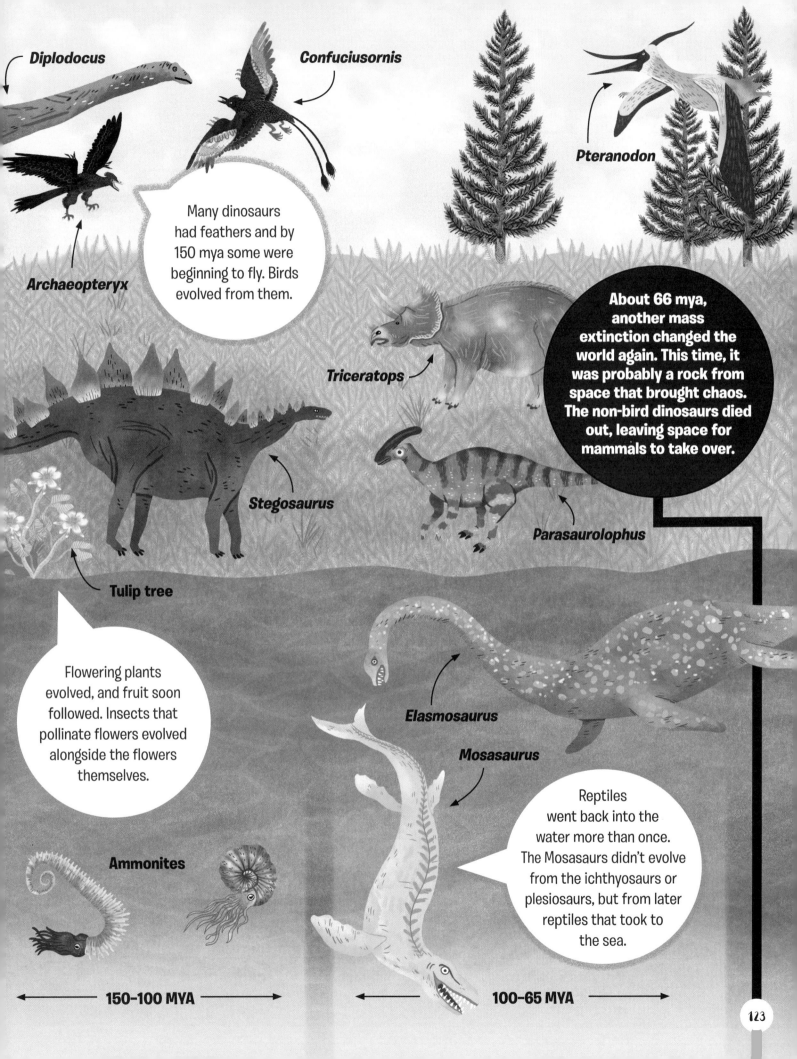

Diplodocus

Confuciusornis

Pteranodon

Many dinosaurs had feathers and by 150 mya some were beginning to fly. Birds evolved from them.

Archaeopteryx

About 66 mya, another mass extinction changed the world again. This time, it was probably a rock from space that brought chaos. The non-bird dinosaurs died out, leaving space for mammals to take over.

Triceratops

Stegosaurus

Parasaurolophus

Tulip tree

Flowering plants evolved, and fruit soon followed. Insects that pollinate flowers evolved alongside the flowers themselves.

Elasmosaurus

Mosasaurus

Ammonites

Reptiles went back into the water more than once. The Mosasaurs didn't evolve from the ichthyosaurs or plesiosaurs, but from later reptiles that took to the sea.

← 150–100 MYA →

← 100–65 MYA →

Argentavis

Some mammals and birds have grown to enormous sizes—just as dinosaurs did before. **Argentavis** was the largest bird ever, with a wingspan of 6 m (20 ft).

Scarlet macaw

Blue morpho Butterfly

Honeybee

Woolly mammoth

Horse

Humans

Humans have killed off many other species in many ways, through hunting, changing the landscape, and now through causing climate change.

Human beings are the first animals to evolve complex technologies, social structures, and languages.

Sahelanthropus

Blue whale

The evolutionary ancestors of humans were apes—the line that split off to produce humans started about 7 mya.

Evolution has not ended. Organisms will continue to change and adapt. Today, they face challenges from the way humans are changing the world. Some scientists think a new extinction event has already begun, this time caused by people. It's impossible to tell what the outcome will be.

Manta ray

← **20–1 MYA** →   ← **1 MYA–NOW** →

GLOSSARY

algae   Water plants of many different types.

amphibian   An animal that lives in the water during it's egg and young stage, but which breathes air when it changes to its adult form.

ancestor   An individual further back in an organism's family tree. Your great-great grandparents are some of your ancestors.

archaea   A very simple single-celled organism.

archosaur   A group of early reptiles including the ancestors of crocodiles and dinosaurs.

arthropod   An animal with a hard, jointed outer skeleton. Examples include crabs, spiders, insects, and millipedes.

asexual reproduction   Reproduction that makes an exact copy of a single parent organism.

asteroid   A rock moving through space.

atmosphere   The layer of gases around a planet (the air around Earth).

atom   The smallest particle of matter.

bya   Billion years ago.

cartilage   A tough material that stiffens parts of an animal's body but is not as hard as bone. The hard part of your nose is made of cartilage.

cell   The smallest building blocks of all living organisms. All organisms have at least one cell.

chromosome   A long molecule of DNA which is divided into genes carrying genetic information about an organism.

climate   Broad patterns of weather, such as the average temperature over a period of years.

common ancestor   An organism from which two or more different organisms have descended.

continent   A large landmass.

convergent evolution   The development of similar features in organisms that have evolved separately. Fins have evolved in both fish and in marine mammals such as dolphins.

crocodilian   An animal with crocodile-like features.

cyanobacteria   Simple bacteria that photosynthesize.

decomposer   An organism that breaks down waste. Microbes in the soil and worms work as decomposers to break down dead animal and plant material.

diversity   Variety in living things.

DNA   A chemical that carries genetic information in the form of a code made up of molecules in a meaningful sequence.

domesticated   An animal kept and changed by people. Dogs and horses have both been domesticated.

environment   The area and conditions where something lives.

equator   An imaginary line around the middle of the Earth, halfway between the North and South Poles.

eukaryotic cell   A complex cell with a nucleus and extra structures called organelles that do particular tasks.

evolution   The process by which organisms change over time, adapting to suit the conditions in which they live.

exoskeleton   An animal's hard outside shell.

extinct   Describes an animal or plant that has disappeared forever.

fishapod   Animal at an evolutionary stage between a fish and a land-going animal with legs.

fossil   The remains or trace of an organism preserved in rock.

fungi   Organisms such as mushrooms, toadstools, and yeast.

gene   A segment of a DNA molecule that holds a chemical code to control an inherited feature (such as a yellow beak, or pointy ears).

generation   A tier in a family tree; you are of one generation, your parents of another, and your grandparents of another generation.

genetic   Relating to inherited characteristics.

gill   Part of the body of a fish that takes air from the water.

global warming   Rising average temperature of the world over a period of years.

imprint   Shape pressed into a surface, such as a footprint in mud.

inherit   Have a feature passed on from a parent.

larva, larval stage   An immature (young) stage in an animal's development when it doesn't have the same body structure as an adult. Caterpillars and tadpoles are both larvae.

lichen A fungus and algae or cyanobacteria living together like a single organism.

lobe-finned fish Bony fish with fins attached to the body by a fleshy stalk.

mammal A warm-blooded animal with fur that gives birth to live young.

marsupial Mammal that gives birth to its young when they are only partly developed, then keeps them in a pouch as they grow.

mass extinction event A period when a very large number of species all die out at the same time.

membrane Thin, flexible sheet of cells that acts as a barrier or boundary in an organism.

microbe A very tiny organism that can only be seen with a microscope.

molecule Group of two or more atoms fixed together.

mutation Error in copying genetic information, so that an organism is different from both its parents in some way.

mya Million years ago.

nucleus Central part of a cell where genetic information is stored.

organism A living thing.

ornithischian A type of dinosaur with hips shaped like those of a bird. They ate plants, and many had a mouth like a beak. *Triceratops* and *Stegosaurus* were ornithischian dinosaurs.

photosynthesize Use energy from sunlight to make a sugar (glucose) from water and the gas carbon dioxide.

predator An animal that hunts other animals to eat them.

primate A type of mammal that includes monkeys, apes, and humans.

prokaryotic cell A simple cell with no nucleus or other structures separated by membranes.

radiation Energy that travels in waves through space.

reproduce To make more organisms of the same type.

reptile An animal that lays eggs on land that hatch into young that look the same as the adults. Most reptiles are cold-blooded (they can't control their body temperature from within) and have scales or plates over their skin.

rodent An animal like a rat, mouse, or squirrel that has a pair of sharp teeth that continue to grow through all of its life, enabling it to gnaw its food.

sauropod Type of dinosaur that usually walked on all fours and had a long neck and tail. *Diplodocus* was a sauropod.

scavenger An animal that eats food it finds lying around, often including other animals that have died.

sediment Mud, sand, and plant material that settles in a river, a pond, or the sea.

sexual reproduction Reproduction that takes genetic information from two parents to make a new organism with a mix of the features of both.

snout The sticking-out nose and mouth of an animal (a dog has a snout; a person doesn't).

soot Tiny black particles from smoky fires.

species Distinct type of organism, such as a Bengal tiger or snowy owl.

sperm Sex cell produced by a male organism that combines with a female's egg cell to produce a new organism in sexual reproduction.

streamlined A smooth shape that moves easily through water or air.

synapsid Type of animal, including mammals and some early reptiles, with a particular skull shape (there is a hole behind each eye in the skull).

theropod A type of dinosaur with strong back legs that usually walked upright and ate meat. *Tyrannosaurus rex* was a theropod.

undergrowth Low-growing plants beneath the trees of a forest.

vegetation Growing plants.

volcano Opening in the Earth's surface through which hot, molten rock can pour or leak.

vulnerable At risk of harm.

wildfire A fire that burns through areas of woodland or grassland.

INDEX